T0317540

APPLICATIONS OF HIGH TEMPERATURE SUPERCONDUCTORS TO ELECTRIC POWER EQUIPMENT

APPLICATIONS OF HIGH TEMPERATURE SUPERCONDUCTORS TO ELECTRIC POWER EQUIPMENT

Swarn Singh Kalsi

IEEE PRESS

A JOHN WILEY & SONS, INC., PUBLICATION

Published by John Wiley & Sons, Inc., Hoboken, New Jersey.
Published simultaneously in Canada.

For general information on our other products and services or for technical support, please contact our Customer Care Department within the United States at (800) 762-2974, outside the United States at (317) 572-3993 or fax (317) 572-4002.

Wiley also publishes its books in a variety of electronic formats. Some content that appears in print may not be available in electronic formats. For more information about Wiley products, visit our web site at www.wiley.com.

Library of Congress Cataloging-in-Publication Data:

Kalsi, Swarn.
  Applications of high temperature superconductors to electric power equipment / Swarn S. Kalsi.
    p. cm.
  Includes bibliographical references and index.
  ISBN 978-0-470-16768-7 (cloth : alk. paper)
  1. Electric machinery–Materials. 2. Electric power systems–Equipment and supplies.
3. High temperature superconductors–Industrial applications. I. Title.
  TK2391.K185 2010
  621.31′042–dc22

                                                                      2010010789

Printed in Singapore.

10 9 8 7 6 5 4 3 2 1

To my wife
Kuldeep
My lifelong companion and aspirant

CONTENTS

PREFACE

On the urging of many colleagues, I undertook this book project with the objective of providing a reference source for designing power equipment with the high temperature superconductor (HTS) developed in the late 1980s. The HTS technology is still in infancy, and both the conductor and its applications are still evolving. The design and analysis approaches discussed are based on experience gained and lessons learned from many coworkers at General Electric Global Research, Northrop Grumman, and American Superconductors over a period of more than 35 years. My understanding of superconductor applications was expanded tremendously during extended assignments at the national laboratories (Oak Ridge National Laboratory and Brookhaven National Laboratory) through association with fusion and accelerator projects. In writing this book, I claim no credit for the original inventions or for anything more than a small part in their subsequent development. The book is merely an attempt to provide a reasonably organized account of the fundamental principles of various power equipment, the basics of the design methodology with example designs, a description of prototypes constructed, and encouragement for readers working to further the HTS technology.

The book presumes a familiarity with the fundamentals of design and analysis of conventional power equipment like motors and generators, transformers, power cables, and electromagnets. The intended audience for the book is electrical and mechanical engineers in the power industry, government laboratories, and students at the senior/ graduate level in universities. Below is a description of topics covered in each chapter.

Chapter 1 contains the introduction to HTS technology, HTS applications to power equipment, and the price goals necessary for the success of this technology.

Chapter 2 provides information on the state of the art of HTS technology. It covers the most popular HTS applications, the characteristic data useful for designing superconducting magnets, and a simple design of an HTS magnet for illustrating the design process.

Chapter 3 describes different types of refrigerators that are necessary to keep HTS coils at their designed cryogenic temperatures between 4 and 80 K. It also includes an introduction to the gases necessary to create a cryogenic environment and to designing cryostats (or enclosures) to maintain superconductors in the required cryogenic environment. An example cryostat design illustrates the design process, with an emphasis on the key design drivers.

Chapter 4 is the longest and presents the basic analysis equations for air-core (no-iron in the magnetic circuit) rotating AC machines. An HTS rotating machine design and analysis are used to illustrate the application of these equations. An example generator design, and many prototype motors and generators already built and tested, are described.

Chapter 5 describes DC homopolar machines, which employ a stationary HTS magnet for excitation and rotary DC armature operating at room temperature. The basic design and analysis equations are presented along with the tested prototypes. These machines have a great potential for application as ship drives.

Chapter 6 describes AC switched reluctance machines, which have both excitation and AC armature windings in the stationary frame. These machines, having the same terminal characteristics as traditional field-wound synchronous machines, are usually built for high-frequency, high-speed operations. A description of a tested prototype is included.

Chapter 7 describes the design and analysis of HTS transformers. Transformers are the most widely used equipment in the power industry, and the HTS transformer technology is ready for commercial exploitation at this time. An example transformer design is included.

Chapter 8 describes various HTS fault current limiter (FCL) concepts. These FCLs have a highly desirable feature of limiting the fault current and then resetting automatically without human intervention. The design and analysis processes for most types of HTS FCLs are included along with the description of the built hardware.

Chapter 9 describes design and analysis for high-capacity HTS power cables. The increase in power demand in inner cities, along with the lack of available space for new cables, is providing incentives for developing highly power dense and thermally neutral HTS power cables. The design and analysis for most types of HTS cables are discussed, including example designs and a description of built cable systems.

Chapter 10 describes principles of magnetic levitation employed in the two most common types of Maglev train systems. Superconducting magnet designs for levitation and propulsion of these trains are included.

Chapter 11 describes many low-field and high-field magnets built with HTS conductors. These magnets have been built for operation at various temperatures, ranging from 4 to 80 K. Smaller size and weight, higher efficiency, higher field, and better thermal stability than the low temperature superconductor (LTS) magnets are some of the attractive attributes of the HTS magnets. A description of some of the prototypes built around the world is included.

I hope that the theory and design presented in this book will help engineers in manufacturing and user communities to appreciate the benefits that the HTS technology can provide to electric power equipment. In light of the current emphasis on renewal of the electric grid, with intention of improving grid reliability and raising efficiency of power transfer, as well as higher efficiency of individual equipment, the HTS technology can play a very significant role. The adaptation of new technologies, like HTS, is usually difficult because the first priority of a user is to minimize risk by not opting for technologies with little reliability and availability data. I urge the current and future engineers to keep developing the HTS technology as it has the potential for solving many vexing problems of the present electric grid and power equipment employed in electric utilities, industry, and ship systems. Currently the HTS technology is at the same stage of development as the electric copper technology was a century ago, and over the next couple of decades, many barriers appearing impregnable will be breached. New HTS conductors and applications will emerge and the current HTS technology will seem primitive.

ACKNOWLEDGMENTS

Many engineers and physicists have contributed to this book through the lessons they have taught me. Their work is referenced throughout the book. However, I would like to acknowledge the following colleagues who helped tremendously by reading many drafts and making excellent suggestions and corrections: Dr. Alex Malozemoff , Peter Win (AMSC), Dr. Bill Hassenzahl (Advance Energy), Dr. Trifon Laskaris and Dr. Kiruba Sivasubramaniam (GE Global Research), Clive Lewis (Converteam), Howard Stevens and Roy Dunnington (BMT Syntek Technologies), Neal Sodengaard (US Navy), David Lindsay (Southwire), Drew Hazelton and Juan Llambes (SuperPower), Dr. Joachim Bock and Robert Dommerque (Nexans Superconductors), Dr. Larry Masur and Dr. Jurgen Keller (Zenergy Power), Dr. Mischa Steurer (Center for Advanced Power Systems, Florida State University), Sam Mehta (Waukesha Transformers), Dr. Alan Wolsky (Argonne National Lab.), Bill Schwenterly and Dr. Jonathan Demko (Oak Ridge National Lab.), Prof. Mitsuru Izumi (Tokyo University of Marine Sciences and Technology), Dr. Robert Buckley (Industrial Research, NZ), Prof. Sheppard Salon and Kent Davy (Rensselaer Polytechnic Institute), many other who provided constructive comments. Last, I thank my wife for bearing my absence for countless hours while I worked on the book.

Princeton, NJ   Swarn S. Kalsi
January 2011

ABBREVIATIONS

Symbol	Meaning
NbTi	Niobium-Titanium
Nb_3Sn	Niobium-Tin
MgB_2	Magnesium Diboride
BSCCO-2212	$Bi_2Sr_2CaCu_2O_8$
BSCC0-2223	$(Bi,Pb)_2Sr_2Ca_2Cu_3O_{10}$
YBCO-123	$YBa_2Cu_3O_7$
SMES	Superconducting Magnetic Energy Storage
HTS	High Temperature Superconductor
FCL	Fault Current Limiter
LTS	Low Temperature Superconductor
B	Magnetic Field
T	Tesla—magnetic field unit
J	Current density
J_c	Critical current density
J_e	Engineering current density
K	Kelvin—temperature unit
OPIT	Oxide-Powder-In-Tube
1G	BSCCO-2223 conductors made with OPIT process
2G	YBCO-123 coated conductors
kAm	kilo-ampere current carried in 1-m length of a wire
H_c	Critical magnetic field
T_c	Critical temperature
MFC	Multi-filamentary composite
MRI	Magnetic Resonance Imaging
SEI	Sumitomo Electric Industries, Japan
AMSC	American Superconductor Corporation, MA, USA
SP	SuperPower, Intermagnetic General Corp., NY, USA
1G	First generation HTS
2G	Second generation HTS

Symbol	Meaning
SCS	Surround copper stabilizer
MOD	Metal organic deposition
RABiTS	Rolling assisted biaxially textured substrate
PLD	Pulse laser deposition
IBAD	Ion beam assisted deposition
LMO	$LaMnO_3$
CTFF	Continuous tube forming and filling
YSZ	Yttria-stabilized zirconia
He	Helium
Ne	Neon
H_2	Hydrogen
N_2	Nitrogen
O_2	Oxygen
LHe	Liquid helium
LN_2	Liquid nitrogen
LNe	Liquid neon
VPI	Vacuum pressure impregnation

1

INTRODUCTION

High temperature superconductor (HTS) materials, discovered in the 1986, are now commercially available worldwide. Two categories of applications have emerged: (1) low-field applications at 77 K achieved with liquid nitrogen, and (2) high-field applications at >25 K achieved with cryogenic refrigerators. The promise of low-cost HTS conductors coupled with reasonably priced refrigeration systems has encouraged application of this technology to a variety of magnets and power equipment. Many prototypes have been constructed for electric power applications such as motors and generators, transformers, power transmission cables, fault current limiters, Maglev trains, and magnets for applied physics research.

According to the US Department of Energy, motors account for three-quarters of all energy consumed by the domestic manufacturing sector and use over half of the total electric energy generated in America. Large electric motors, those greater than 1000 horsepower, consume over a third of the total generated electric energy, and three-quarters of these motors are suited to utilize HTS technology. With some minor exceptions, nearly all cruise ships today employ electrical propulsion, and many other types of commercial vessels and warships are adopting marine motors as their primary source of motive power.

Applications of High Temperature Superconductors to Electric Power Equipment,
by Swarn Singh Kalsi
Copyright © 2011 Institute of Electrical and Electronics Engineers

Superconducting technologies could also benefit ancillary equipment for smooth operation of the electric grid; examples of such applications are superconducting electromagnetic energy storage (SMES) and fault current limiters (FCLs). In addition to improving system efficiency, superconductors reduce size and weight of equipment. Transformers are the prime candidates for employment of the HTS technology because they are the most widely used equipment in an electric grid. Most of the power is generated as AC at relatively low voltages and is utilized at even lower voltages, but the bulk of power is transmitted from points of generation toward points of consumption at very high voltages. Transformers convert electric power from one voltage level to another.

The electric grid inevitably experiences extreme natural events and faults. Fault current limiters, such as fuses, are employed to limit the fault current and allow an electric grid to keep operating. However, fuses require manual replacement after a fault. The HTS fault current limiters are self-acting and resetting devices that allow the grid to recover quickly following a fault.

Power cable built with HTS wire could carry several times more power than the conventional copper cables of similar physical size. The space freed by use of HTS cable is available for enhancing power transmission or other applications.

Large electrical magnets are employed for a variety of industrial, research, and military applications. The manufacture of such magnets with HTS materials is looking attractive. The HTS magnets possess attractive features such as smaller size and weight, higher efficiency, higher fields, better stability, longer life, and easier cooling.

The four main HTS materials are BSCCO-2223, YBCO-123, BSCCO-2212, and MgB_2. However, only BSCCO-2223 and YBCO-123 wires capable of operation in the temperature range of 20 to 70 K have achieved widespread application for manufacturing practical electric power equipment. The YBCO-123 in form of coated conductors has also advanced significantly to provide current density capability suitable for application in practical devices. Roebel cables, made from the coated conductors, could carry currents in kA range with minimal losses. The BSCCO-2212 wire has also been utilized for building insert magnets for operation at 4 K in high background fields. This material in form of bulk rods is also employed for building fault current limiters operating in liquid nitrogen baths. MgB_2 conductors are also being fabricated, but their current-carrying capability is still lower than the other HTS materials.

Despite higher field and higher operating temperature capabilities than the low-temperature superconductor (LTS) materials (NbTi and Nb_3Sn), the MRI, accelerator, and fusion magnet communities have not adopted the HTS materials so far. These applications employing LTS materials are purely DC magnets operating at around 4 K and are very sensitive to even a small thermal energy injection that could raise local temperature sufficiently to drive the magnet into quench. Very low heat capacity of materials at low temperatures (around 4 K) causes this phenomenon. Then again, materials in HTS conductors operating at about 30 K have heat capacity hundreds of times higher than that at 4 K. Thus any local energy injection causes a much smaller temperature rise. HTS materials also transition slowly to their normal state because of a low N-value.* These two factors, enable the HTS magnet to operate successfully in the presence of significant local heating.

HTS materials have proved to be successful in many applications where LTS materials have been unsuccessful especially for industrial applications. HTS conductors operating at liquid nitrogen temperature are employed for building power cables, fault current limiters (FCL), and transformers for applications in AC electric grid. Numerous large projects in these areas have been successfully built. The HTS conductors are also used for building high field DC coils for excitation poles of large AC synchronous motors and generators. These coils are employed on the rotor and are cooled with a stationary refrigerator employing thermo-siphon or gaseous helium loops. The cooling is accomplished with an interface gas like He or neon, which transports thermal load from coils to the stationary refrigerator. A few examples of many large motors and generators successfully built include 8-MVA and 12-MVAR reactive (MVAR) synchronous condensers, the 4-MW, two-pole generator for marine application, 5-MW and 36.5-MW high-torque ship propulsion motors, and 1000-hp and 5000-hp industrial motors.

The largest potential market for HTS conductors lies in the electric power arena and involves power transmission cables, high-power industrial and ship propulsion motors, utility generators, synchronous condensers, fault current limiters, and transformers. The Northeast blackout of August 14, 2003, provided incentive to accelerate development of such power equipment utilizing HTS wires.

This book reviews key properties of HTS materials and their cooling systems and discusses their applications in power devices such as

---

*N is the exponent of the V-I curve for the HTS wire.

rotating machines, transformers, power cables, fault current limiters, Maglev trains, and magnet systems. Example device designs are included to allow the reader to size a device for his/her application. Ample references are provided for exploring a given application in details.

So far the high cost of HTS materials has inhibited commercial adoptation of HTS power devices. Estimates of the maximum acceptable price for different applications range from $1 to $100 per kiloamp-meter, where kiloamp (kA) refers to the operating current level. In particular, NbTi wire typically sells for $1/kAm but is limited to operation in the liquid helium temperature range. HTS conductors operating at 20 to 77 K are expected to be economical for some applications in the $10 to $100/kAm range. In power equipment, copper wires typically operate in the range of $15 to $25/kAm; this sets a benchmark for superconducting wire. Thus HTS materials must compete against copper in electric power technology where cost pressures are very significant.

Many synchronous machines prototypes (both generator and motor) were developed worldwide between 1995 and 2010. These machines included slow-speed machines (100—200 RPM) for ship propulsion applications and high-speed machines (1800 and 3600 RPM). Since most of these prototype machines employed expensive BSCCO-2223 HTS material, it was not possible to transition them into products. However, currently available YBCO-123 coated conductors promise low-cost and higher current capability. This is encouraging renewed interest in developing synchronous generators for central power stations and wind farms.

Underground power transmission HTS cables have been amply demonstrated with both BSCCO-2223 and YBCO-123 coated conductors. Many prototypes are currently in operation throughout the world at voltage ratings ranging from 11 to 138 kV. Although these prototypes demonstrate benefits of the HTS technology for the users, high capital cost and reliability issues are inhibiting their adoption.

Today a number of fault current limiter (FCL) projects are underway. Most employ YBCO-123 in tape or bulk form. Some prototypes are already operating in electric grids. FCLs are solving electric utility problems that are difficult to address with currently available technologies. However, the success of FCLs will be determined by the capital cost and reliability considerations.

Many power transformers were also prototyped between 1995 and 2005 all over the world. Most employed BSCCO-2223, which was not found suitable for this application because of high AC losses. New projects are being initiated now with YBCO-123 coated conductors, as these promise to reduce losses in the acceptable range. These conductors are being considered as wide tapes or in form of Roebel cables. Although the HTS transformers offer higher efficiency and lower weight, their future will be determined by the economic viability and reliability in operation.

To encourage future customers to adopt the HTS technology, it is essential that standards be developed on par with those for similar conventional equipment. HTS equipment must satisfy all requirements imposed by current industry standards (CIGRE, IEEE, NEMA, VAMAS, etc.) on the existing conventional equipment. If a certain requirement cannot be met by HTS equipment, then alternate arrangements must be made to ensure that the performance and reliability of new equipment are on par with currently available conventional equipment. The standards activity must be started now. Presently, IEEE and CIGRE are initiating standards development activity for HTS rotating machines and FCLs.

Although many niche applications of the HTS technology are feasible, only a widespread adoption will lower the cost of HTS materials and the cooling systems. This is a typical chicken-and-egg problem. Most applications require lower cost HTS conductors and cooling systems, but developers require more demand (orders) for reducing cost. By the 2020s some economic products are likely to emerge.

2

HTS SUPERCONDUCTORS

2.1 INTRODUCTION

Some materials become superconductors—meaning they are capable of conducting electricity with zero resistance—when cooled to very low temperatures. These materials can be applied to a variety of electrical equipment with significant benefits. For example, in the US electric grid, nearly 7% of the total power is wasted in transmission from point of generation to the users. A similar amount is lost at points of generation and distribution. This is a significant loss with adverse impact on natural resources and the environment. System efficiency could be improved significantly by employment of superconductors in power-generating and distribution equipment. Power technologies that can benefit from the application of superconductors include large motors and generators, transformers, and power transmission cables. A related application is the magnetically levitated train (Maglev). Superconducting technologies could also benefit ancillary equipment for smooth operation of the electric grid—examples of such applications are superconducting electromagnetic energy storage (SMES) and fault current limiters (FCLs). In addition to improving system efficiency, superconductors reduce the size and weight of equipment. Compact and lighter motors and

Applications of High Temperature Superconductors to Electric Power Equipment, by Swarn Singh Kalsi
Copyright © 2011 Institute of Electrical and Electronics Engineers

generators are welcome for applications on ships and aircraft and on both Maglev and conventional trains.

Superconductors discovered early on were pure metals like mercury, lead, and tin, which had practically zero current-carrying capability in presence of high magnetic fields. However, in the late-1950s and early-1960s, new class of high-field alloys like niobium-titanium (Nb-Ti) and compounds such as niobium-tin (Nb_3Sn) were discovered. They are employed in commercial devices such as MRI magnets and a variety of high-field magnet systems. These materials must be cooled to about 4 K, where the cost of the cooling system and its reliability are challenging. As a result these materials have not found wider application.

The discovery in 1986 of superconductivity in the range of 35 to 40 K and the subsequent increase in critical temperatures to 110 K prompted their consideration for a variety of electric power equipment. These materials are classified as high temperature superconductors (HTS). They can be cooled with liquid nitrogen for applications in low magnetic fields. They can be used with off-the-shelf cryocoolers for high magnetic fields in the range of 30 to 40 K. Prerequisite to wide application of this technology is the development of innovative techniques for improving the performance and reducing the production cost of HTS materials and the necessary cooling system. In this chapter a brief history of superconductivity is reviewed, followed by discussion of a variety of available high temperature superconductors. Primary focus is on BSCCO-2212, BSCCO-2223, YBCO-123, and MgB_2 materials. These superconductors offer the promise of increased efficiency and lower operating cost for the electrical industry. A much more detailed treatise on superconductors is provided elsewhere [1]. An example magnet design is presented later in this chapter for illustrating superconducting magnet design process.

2.2 HTS BACKGROUND AND NOMENCLATURE

2.2.1 Background

Kamerlingh Onnes discovered superconductivity 1911, which followed liquefaction of helium at his institute in 1908. This relationship between superconductivity and cooling system is discussed by Blaugher [2]. Two commercially available, low temperature superconductors (LTS) are Nb-Ti and Nb_3Sn. Although their critical temperatures are 9 and 18 K, respectively, for most applications they must operate below 5 K, which often requires expensive refrigerators that are difficult to operate in an

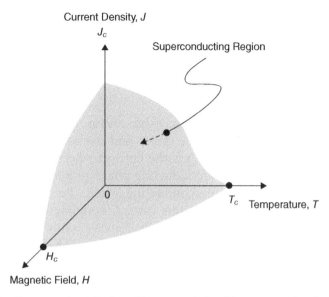

Figure 2.1   Superconductivity in a 3D space defined by current density, temperature, and magnetic field

industrial environment. However, the new ceramic HTS materials require less cooling and could operate at temperatures higher than 77 K, the equilibrium temperature of liquid nitrogen at atmospheric pressure, for low-field applications and 30 to 40 K for high-field applications. The room-temperature electric power required to operate a 77 K refrigerator is less than 1/10th that required for a 4 K refrigerator.

All superconductors must operate within a regime bounded by three inter-related critical meters, current density, operating temperature, and magnetic field as shown in Figure 2.1. The highest temperature at which a material possesses no electrical resistivity (exhibits superconductivity) is called its critical temperature (T_c). The upper limit on its current-carrying capability is called the critical current density (J_c) and critical magnetic field (H_c) is that above which it ceases to be a superconductor. The superconductor will revert to its normal state if any of these limits is exceeded. Since their discovery in 1986, HTS materials are beginning to see applications in prototype motors and generators, power transmission cables, transformers, fault current limiters, and magnets for Maglev trains.

The most common HTS conductors are BSCCO-2212, BSCCO-2223, YBCO-123, and MgB_2. These are manufactured by a variety of techniques. The following sections provide a brief description of each conductor, manufacturing processes, and limitations of its use. Readers interested in learning more about these materials should refer to a paper by Scanlan et al. [3] that also includes an extensive list of references.

2.2.2 Nomenclature

Before proceeding further, it may be useful to define quantities used by superconductor manufacturers and users. The ideal physical definition of critical current (I_c) is the current where a material has a phase transition from a superconducting phase to a nonsuperconducting phase. For practical superconducting wires, the transition is not infinitely sharp but gradual. In this case, I_c is defined as the current where the voltage drop across the wire becomes greater than a specified electric field (E_c), usually $1\,\mu\text{V/cm}$ in self-field at $77\,\text{K}$ (liquid nitrogen temperature). The trend, however, at the operating field of the device is to use the more restrictive value of $0.1\,\mu\text{V/cm}$, which is typically used for commercial low-temperature superconductors. The self-field is the magnetic field that is created by a finite current flowing in the wire. During critical current measurements, the self-field is the magnetic field that is induced in a straight piece of wire that is being measured.

A superconductor wire is characterized by a *V-I* curve, or the "voltage drop along the length of wire as a function of its current." A generic *V-I* curve is shown in Figure 2.2. The industry accepted practice is to define the superconductor's critical current (I_c) as the current that produces a voltage drop of $1\,\mu\text{V/cm}$, as shown in the figure. An *N*-value, the characteristic of a given superconductor, describes the relationship of the voltage drop across the wire to the applied current.

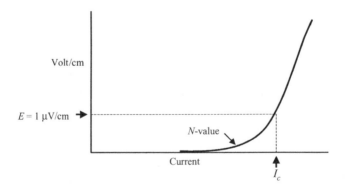

Figure 2.2 *V-I* curve for a superconductor

For the transition from zero resistance (zero voltage drop) to a finite resistance (finite voltage drop), the *V-I* curve of superconducting wires can be fitted with the power law given in equation (2.1).

$$E(I) = E_c \left(\frac{I}{I_c} \right)^N ,$$ (2.1)

where $E(I)$ is the longitudinal voltage drop across the superconductor, E_c is the electric field criterion at $1.0\,\mu V/cm$, I is the current in the conductor, I_c is the critical current, and N is the exponent. A higher N-value produces a sharper transition in the *V-I* curve and indicates the quality of a superconductor. A high-quality superconductor will have high N-value. The loss per centimeter length is equal to the product of $E(I)$ and I. Generally, I is a fraction of I_c and a higher N-value translates into lower loss.

Generally, superconducting materials are compared on the basis critical current density (J_c), which is the critical current of a superconductor divided by its cross-sectional area. However, in the design of superconducting devices the important quantity is the engineering critical current density (J_e), which is the critical current of the wire divided by the cross-sectional area of the entire wire, including superconductor and normal conductor. The value of J_e is important in the manufacture of practical devices as it determines coil cross section, which includes additional components of the windings such as supports and insulation.

2.3 BSCCO-2212 CONDUCTORS

BSCCO-2212 has critical temperature (T_c) of about $90\,K$ and is interesting primarily for its high-field properties in the temperature range between 4.2 and $20\,K$. This material has a very low current-carrying capacity in a magnetic field at $77\,K$. It was the first HTS material used for making superconducting wires and is very versatile. BSCCO-2212 can be made in the form of round wire, flat tape, cast bulk rod, and blocks and in a large variety of shapes and dimensions. Its ability to melt and recrystallize allows flexibility in conductor geometry. The critical current in the round wires shows no anisotropy with respect to an applied field. Several successful methods of achieving high current density conductors have been demonstrated, including powder-in-tube,

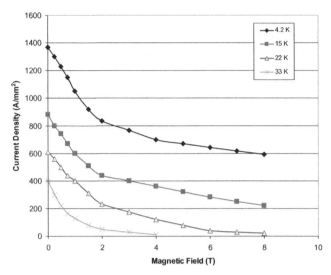

Figure 2.3 Critical current (J_c) for BSCCO-2212 as a function of magnetic field (B) at different temperatures

dip-coating, tape casting, electrophoretic coating, and spray coating. Conductors of this material have been made in long lengths of wire with uniform properties without breakage. This versatility of BSCCO-2212 principally provides a potential for wide engineering use where the operating temperature is lower than 20 K.

The critical current density (J_c) as a function of magnetic field (B) is given in Figure 2.3 for different temperatures for an Oxford Superconductor Technologies (OST) BSCCO-2212 conductor measured by Barzi [4]. The I_c was measured for an OST 0.7-mm diameter round conductor. These data were converted to J_c by dividing the measured I_c by the conductor's cross section as shown in Figure 2.3. The J_c drops dramatically at higher temperatures. Marken [5,6] has published extensively on the manufacturing and use of BSCCO-2212 conductors.

Thin tapes of BSCCO-2212 have achieved [3] current densities of 7100 A/mm in self-field and 3500 A/mm in 10 T parallel to the tape plane at 4.2 K. This has enabled this conductor's application in high magnetic fields, and so the inserted coils have been used for several record-setting magnetic field levels generated by superconducting magnets. At present, BSCCO-2212 holds the record for producing the highest field with a superconducting magnet. Multifilament tapes have been made for application to high-field magnets, and Rutherford cables for accelerator magnets and other high current superconducting magnets. This wire can be used as a direct substitute for LTS wires in Rutherford-type cables in accelerator magnets.

As stated earlier, performance of BSCCO-2212 is excellent below 20 K, but at temperatures higher than 20 K its performance is limited. It was eclipsed by BSCCO-2223 wires, as discussed in the next section.

2.4 BSCCO-2223 OPIT WIRES

BSCCO-2223 HTS has a critical temperature of about 110 K, which is about 20 K higher than BSCCO-2212. The oxide-powder-in-tube (OPIT) method has been used for manufacturing flexible multifilamentary composite conductor by many manufacturers [7] in what is named first-generation (1G) wire. Highest average I_c of 150 A in self-field at 77 K was achieved in 4-mm-wide tapes. The OPIT manufacturing process was successfully implemented in industrial production and achieved continuous tape lengths of several hundred meters while simultaneously improving tape price and capacity. These 1G HTS tapes were a commercial reality, and their electrical, mechanical, and thermal performance capabilities allowed introduction into electric machinery, power cable, and other applications. However, the lowest cost of this material was still roughly 100 \$/kAm in 2005, while the cost goal for wide industrial use is believed to be \$10/kAm. Sumitomo Electric Industries (SEI) have projected a price of \$20/kAm beyond 2011 (according to their public news release in 2006).

2.4.1 Manufacturing Process

This first-generation HTS wire has a tape shape, typically 0.2×4 mm, consisting of 55 or more tape-shaped filaments, each 10 μm thick and up to 200 μm wide, embedded in a silver alloy matrix. The filaments consist of grains of BSCCO-2223, up to 20 μm long, often arranged in colonies sharing a common axis. For some applications the wire is laminated to stainless steel tapes on either side for enhanced mechanical properties and environmental protection. BSCCO-2223 wire was manufactured in 2004 by a number of companies, including American Superconductor Corporation (Devens, Massachusetts), European Advanced Superconductor GmbH (Hanau, Germany), Innova Superconductor Technology (Beijing, China), Sumitomo Electric Industries Ltd. (Japan), and Trithor GmbH (Rheinbach, Germany). Worldwide capacity has exceeded 1000 km per year.

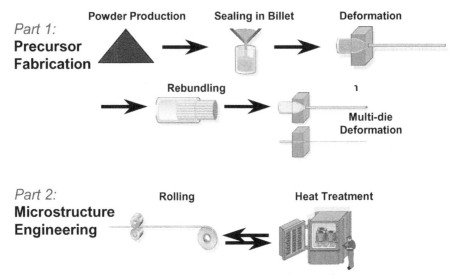

Figure 2.4   Schematic of the OPIT process for manufacturing BSCCO-2223 wire (Courtesy of American Superconductor Corporation)

While precise details of the industrial production process are not public, the basic PIT deformation process is schematized in Figure 2.4. A powder consisting of a mixture of Pb-containing BSCCO-2212, alkaline earth cuprates, copper oxide, and other oxides is initially prepared by an aerosol, spray-dry or simple oxide mixing process; the overall cation stoichiometry is chosen to match that of BSCCO-2223. The powder is packed into a silver tube, which is sealed, evacuated, and drawn through a series of dies, elongating it into a hexagonal rod. These rods are next cut, assembled into a multifilament bundle (55–85 filaments are typical), and inserted into a second silver or silver alloy tube. This tube is sealed, evacuated, and further drawn into a fine round wire, which is subsequently deformed to a tape shape in a rolling mill, in multiple steps of rolling and heat treatment to texture and react the powder inside the wire to form BSCCO-2223. The superconducting fill factor is typically 30% to 40%.

2.4.2   Characteristics—Electrical and Mechanical

Until 2006 three significant sources for the BSCCO-2223 wire were American Superconductor (AMSC, USA), Sumitomo Electric Industries (SEI, Japan), and European High Temperature Superconductor (EHTS, Germany). This section describes characteristics of wires produced by AMSC and Sumitomo—the two

BSCCO-2223 Conductor   Definition of $B_\perp$ and $B_\parallel$

Figure 2.5 Cross section of a typical BSCCO-2223 conductor of 0.2 mm × 4 mm (Courtesy American Superconductor Corporation)

manufacturers of this wire. The basic BSCCO-2223 wire with a rectangular cross-section of 0.2 × 4 mm is shown in Figure 2.5, along with definitions of perpendicular ($B_\perp$) and parallel ($B_\parallel$) magnetic fields with respect to the conductor's cross section. The conductor had typically 85 superconductor filaments embedded in silver alloy matrix. The main technical issue with this wire was that well-annealed silver, which was the state of the conductor sheath after the final thermal process step, was mechanically very weak, having a low yield stress, and it had a large mechanical mismatch to the BSCCO-2223 core. To withstand large Lorentz forces created in magnet applications, doped silver sheaths with higher mechanical strength was considered by some manufacturers. However, American Superconductor created high-strength HTS wire by sandwiching the BSCCO-2223 wire between two stainless steel tapes.

American Superconductor (AMSC) Wire   American Superconductor supplied BSCCO-2223 wires in many configuration; two key configurations (high current density and high strength) are detailed in Table 2.1. The superconductor content in the tape cross section was typically between 30% and 40%. The high current density wire was for use in high-performance coils and magnets, but it had lower mechanical strength (rated at 65 MPa.) The high-strength wire was built for applications where high mechanical strength and tolerance to small bend diameters was desired. The high strength was achieved by sandwiching bare BSCCO-2223 wire between two stainless steel tapes—all held together with a solder. This wire exhibited very high tensile strength—200 Mpa at room temperature and 250 Mpa at 77 K. It had a critical bend diameter of 38 mm. This wire could handle high tensions necessary for producing high-quality coils and cables. These wires were sold in nominal ratings (at 77 K, self-field) of 125 A, 135 A, 145 A, and 155 A.

The critical current (I_c) as a function of field at different temperatures is shown in Figure 2.6 for the 155 A wire—these curves can be linearly scaled for wires of different current ratings. The I_c (data from

Table 2.1   Characteristics of BSCCO-2223 wire from American Superconductor Corporation

Parameters	High Current Density	High Strength
Current density of 155 A wire, A/cm^2	17,200	13,300
Average thickness, mm	0.21–0.23	0.255–0.285
Width, mm	3.9–4.3	4.2–4.4
Minimum double-bend diameter, mm	100	38
Maximum rated tensile stress, MPa		
• Room temperature	65	200
• Liquid nitrogen temperature (77 K)	65	250

Source: Data from American Superconductor website www.amsc.com in 2006.

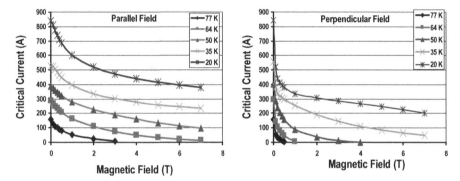

Figure 2.6   Critical current as a function of magnetic field at different temperatures when (*a*) field is llel to the conductor flat side ($B_{\parallel}$) and (*b*) field is perpendicular to the conductor flat side ($B_{\perp}$) for a 155 A wire (Courtesy American Superconductor Corporation)

the AMSC website in 2006) is provided for two magnetic field orientations with respect to the conductor: (1) when the magnetic field is perpendicular to the plane of conductor tape ($B_{\perp}$) and (2) when the magnetic field is llel to the plane of conductor tape ($B_{\parallel}$). The I_c is much lower for $B_{\perp}$ than for $B_{\parallel}$. Thus, during a magnet design, it is necessary to pay attention to this orientation of the field, lest there would be excessive losses in the winding.

American Superconductor ceased manufacture of this wire in mid-2006 and focused on developing YBCO-123 coated conductor.

Sumitomo Electric Industries (SEI) Wire   Sumitomo Electric Industries supplied BSCCO-2223 wire in two configurations (high current density and high strength) in 2006 as detailed in Table 2.2. SEI developed a solid phase method for combined processing of silver and bismuth-based HTS material. Using its innovative process called "con-

Table 2.2   Characteristics of BSCCO-2223 wire from SEI

Parameters	High Current Density	High Strength
Current density of 150 A wire, A/cm^2	15,000	12,000
Average thickness, mm	0.22 ± 0.02	0.22 ± 0.02
Width, mm	4.2 ± 0.2	4.2 ± 0.2
Minimum double-bend diameter, mm	70	50
Maximum rated tensile stress, MPa		
• Room temperature	100	170
• Liquid nitrogen temperature (77 K)	135	210

Source: Data from Sumitomo Electric Industries website www.sei.co.jp in 2007.

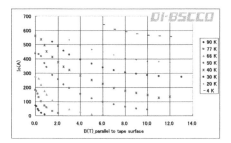 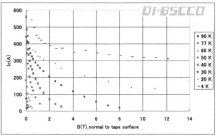

Figure 2.7   Critical current as a function of magnetic field at different temperatures when (left) field is llel to the conductor flat side ($B_{\parallel}$) and (right) field is perpendicular to the conductor flat side ($B_{\perp}$) for a 150 A wire (Courtesy of Sumitomo Electric Industries)

trolled overpressure (CT-OP)," quality and productivity of the wire was improved substantially. The critical current was increased 30% and mechanical strength was improved by more than 50%. The mechanical tensile stress of the SEI high-strength wire was 170 MPa at room temperature and 210 MPa at 77 K. It had a critical bend diameter of 50 mm. The wire density was improved to 100% from conventional 85%. This wire was still commercially available in 2007. These wires were supplied in nominal ratings (at 77 K, self-field) of 110 A, 120 A, 140 A, and 150 A. The performance of SEI wire (taken from their website in 2007) is shown in Figure 2.7 for $B_{\parallel}$ and $B_{\perp}$.

2.5   YBCO-123 COATED CONDUCTORS

In llel to scaling up production for 1G wire, many manufacturers also pursued the development of second-generation (2G) HTS materials based on thin films of an yttrium barium copper oxide YBa$_2$Cu$_3$O$_7$, or

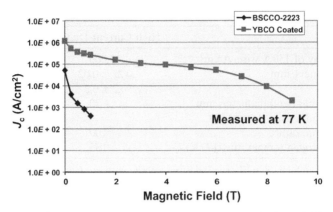

Figure 2.8 Performance comparison between typical 1G and 2G wires

YBCO-123. Because of material characteristics of BSCCO-2223 based HTS material, 1G wires were limited to temperatures lower than 40 K while operating in magnetic fields above 2T. Relative performance of typical 1G (BSCCO-2223) and 2G (YBCO coated) wires is compared in Figure 2.8 at 77 K. As is evident, the 2G wire retains its J_c at much higher magnetic fields than the 1G wire.

The 2G wire has a significantly different architecture compared to 1G wire. Unlike the 1G wire, the 2G wire did not employ noble metal like silver, which is still the main hurdle to achieving low-cost 1G wire. The main benefits of 2G wire are a potential for a two to three times cost reduction due to the higher critical current (I_c), a higher throughput, and a lower manufacturing cost with automation. The cost and availability in long length with high performance were considered to be two key requirements for commercial viability of 2G wires. Also as a form-fit-function replacement for 1G wire, 2G wire may require minimum re-engineering of applications already developed and commercialized using 1G wire. The 2G wire with its greater strain capability was also expected to enlarge the configuration options available to power equipment designers. While many other manufacturers are developing 2G conductors worldwide, the two technologies MOD/RABiTS process adapted by AMSC and the IBAD-MgO process adapted by SuperPower are discussed here because they are most developed and represent the two most common manufacturing processes in use. Competitively priced 2G wire is available from several sources worldwide. This section discusses MOD/RABiTS and IBAD-MgO processes currently in use in the United States.

American Superconductor Corporation's 2G Wire The architecture of 2G YBCO wire, shown in Figure 2.9, comprises multiple coatings on a base material, or a substrate adapted to achieve the highest performance and low-cost superconductor material. AMSC selected the "metal organic deposition (MOD)/rolling assisted biaxially textured substrate (RABiTS)"—or simply "MOD/RABiTS"—process for manufacturing 2G wire. AMSC further chose to laminate a 2G wire with two normal metal tapes to create a "three-ply" architecture—consisting of 2G tape sandwiched between thin copper or stainless steel tapes—that is similar to their commercial 1G HTS wire. It also leveraged technology, manufacturing equipment, and expertise from the company's 1G manufacturing operation. AMSC's 2G wire products, 344 and 348 superconductors, were developed as a drop-in replacements for 1G HTS wire applications for power cable, motor, generator, synchronous condenser, transformers, and fault current limiters, for example, without costly re-design or re-tooling. Cost models for 2G coated conductors, particularly those made by nonvacuum methods, predicted costs for future large-scale production well at $77\,K, 0T$ below $10/kAm, the effective price performance of copper [8].

American Superconductor coated conductor architectures (Figure 2.9) include a flexible substrate, preferably of strong and nonmagnetic or weakly magnetic metal, typically 50-μm thick. On top of the substrate is a multifunctional oxide barrier or buffer layer, typically less

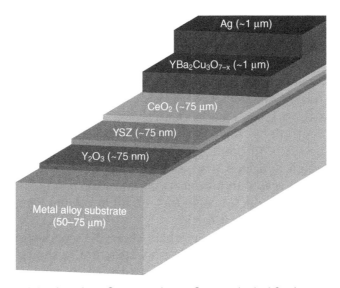

Figure 2.9 American Superconductor Corporation's 2G wire structure

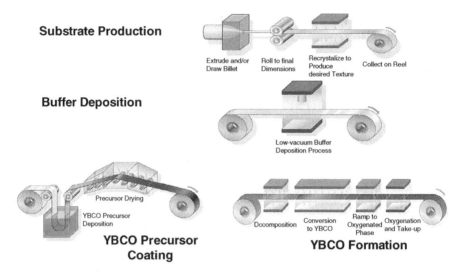

Substrate Production

Extrude and/or Draw Billet — Roll to final Dimensions — Recrystalize to Produce desired Texture — Collect on Reel

Buffer Deposition

Low-vacuum Buffer Deposition Process

Precursor Drying

YBCO Precursor Deposition

YBCO Precursor Coating

Decomposition — Conversion to YBCO — Ramp to Oxygenated Phase — Oxygenation and Take-up

YBCO Formation

Figure 2.10   American Superconductor Corporation's reel-to-reel 2G wire manufacturing process

2G wire   Copper or Steel tape

Figure 2.11   American Superconductor Corporation 344 conductor—2G conductor sandwiched between copper tapes

than 0.5 μm thick, on top of which is the superconducting YBCO layer 1 to 3 μm thick. A protective silver layer of a few μm and a thicker copper protection and stabilization layer completed the conductor [9]. Progress on continuous reel-to-reel processes (shown in Figure 2.10) for making coated conductors progressed rapidly in 2005 to 2006 period. Continuous 10 to 100 m lengths of 1-cm-wide conductor with end-to-end critical currents as high as 270 A/cm width (self-field, 77 K) indicated the success of this multiple-step fabrication process. Uniformity along the length indicated that kilometer-length wires were possible. This process was scaled to producing 40-mm-wide conductors in 2006. The 40-mm-wide conductor could then be slit into 4 mm, 10 mm, or any other width desired by a customer.

American Superconductor markets their 2G conductors as 344 superconductors, which have a three-ply structure as shown in Figure 2.11. The wire's three-ply architecture consists of 2G tape sandwiched between thin copper or stainless steel tapes. The key meters of this

Table 2.3   Characteristics of 2G wire from American Superconductor Corporation (Copper Laminated)

Specification	
Minimum I_c, A	90
Average thickness, mm	0.20 ± 0.02
Width, mm	4.83 ± 0.12
Minimum bend diameter, mm	30
Maximum tensile stress at RT, MPa	150
Maximum tensile strain at 77 K	0.3%
Continuous piece length, m	500

Source: Data from American Superconductor website www.amsuper.com in 2010.

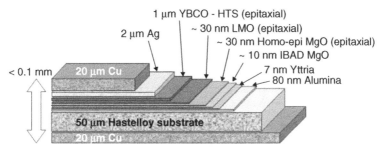

Figure 2.12   Structure of Super Power® 2G HTS wire

conductor are summarized in Table 2.3. Mechanical properties of this 2G wire exceeded the 1G wire.

SuperPower's 2G Wire   SuperPower, a subsidiary of Royal Philips Electronics N.V. adapted the ion beam assisted deposition (IBAD) process for manufacturing 2G conductors. In 2006 SuperPower successfully transferred from low-throughput IBAD-YSZ (Yttria-stabilized zirconia) technology to high-throughput IBAD-MgO technology [10]. A pilot-scale IBAD system and a pilot-scale buffer deposition system, each with capabilities of producing single-piece lengths over 1000 m, were established. The layered structure of the 2G conductor is shown in Figure 2.12. The IBAD-MgO buffer stack includes five layers between the metal substrate and HTS layer in order to make it robust for HTS film deposition. The first-layer alumina mainly serves as diffusion layer to prevent elements of metal substrate from diffusing into other buffer layers and HTS layer during deposition. The second-layer yttria serves as a seed layer to help IBAD-MgO nucleation; the third-layer IBAD-MgO is the key layer, which forms biaxial texture by ion beam assisted

Figure 2.13   SuperPower Helix tape handling

Table 2.4   Characteristics of 2G wire from SuperPower (Copper Stabilized)

Specification	
Minimum I_c, A	80–140
Average thickness, mm	0.1
Width, mm	4
Minimum bend diameter, mm	11
Maximum tensile stress at RT, MPa	>550
Maximum tensile strain at 77 K	0.45%
Continuous piece length, m	Up to 500

Source: Data from SuperPower website www.superpower-inc.com in 2010.

deposition. The fourth-layer homo-epi MgO makes the IBAD-MgO robust and improves the texture, and the fifth layer is the cap layer to provide a good match with the HTS layer. LaMnO$_3$ (LMO), the cap layer, is instrumental in achieving higher I_c. A Helix tape handling approach instead of the wide tape approach (Figure 2.13) was chosen due to its advantages over wide tape approach such as capability to handle longer lengths, shorter process time, and better uniformity across tape width. Both pilot systems were in routine production mode in March 2006.

SuperPower markets 2G wire in four widths: 3, 4, 6 and 12 mm. A surround copper stabilizer (SCS) is applied to completely encase the wire. This stabilizer protects the conductor and produces rounded edges that are advantageous in high-voltage applications. The key parameters of this conductor are summarized in Table 2.4.

The critical current (I_c) of the standard 4-mm-wide 1G and 2G wires over a broad range of temperatures is shown in Figure 2.14. The 2G

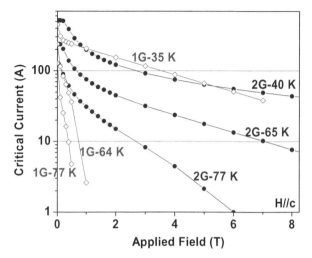

Figure 2.14 Comparison of 1G and 2G wire performances (Courtesy American Superconductor Corporation)

wire has superior performance at all temperatures. The major difference is for applications at 77 K; namely some high field applications could employ 2G wire cooled by liquid nitrogen. This is advantageous, as the cost of refrigeration system will be substantially reduced. In 2010 the performance of 2G conductors were still improving, and the reader is advised to contact wire manufacturers for the latest capability curves for their wires.

2.6 MAGNESIUM DIBORIDE (MgB₂)

In early 2001 the Akimitsu group in Japan discovered that the long ago synthesized compound magnesium diboride (MgB_2) was a hitherto unappreciated 40-K superconductor [11]. Since that discovery, Hyper Tech Research (USA) has been working on making the compound into a high-performing, low-cost superconductor wire. This material essentially has applications at temperatures below 25 K. It can be made by the powder-in-tube (PIT) process, and many groups have made prototype wires using either prereacted (ex situ) MgB_2 powder or mixtures of Mg and B powders, which must be reacted to MgB_2 in situ within the wire [12]. Above about 25 K there is neither a current density nor an upper critical field barrier to the application of MgB_2. This wire could become a credible competitor to LTS-based wires or to BSCCO-based wires used in low-temperature (25 K) applications. An additional quality is that the transition appears to be rather sharp, much more similar to LTS than to HTS. Thus high values and even persistent mode magnets appear feasible, making MgB_2 a potential NMR or MRI magnet conductor. Another advantage of MgB_2 is that the raw material costs of both B and Mg are low, and a commercial conductor is expected to cost several times less than the Nb-based superconductors. The problems of fabricating MgB_2 conductors commercially are being addressed by Hyper Tech Research, Columbus, OH; Columbus Superconductors, Genova, Italy; and Hitachi, Hitachi City, Japan. If costs can be kept low, competition with Nb-base LTS conductors is likely. Competition with either BSCCO-based or YBCO-based 1G and 2G HTS conductors will depend on the progress made in expanding the temperature range for applications requiring high currents in several tesla fields.

Hyper Tech Research developed and patented the continuous tube forming and filling (CTFF) process to make a powder metallurgy-based wire of the MgB_2 superconductor. A conceptual layout of the process is shown in Figure 2.15. Wire is started with a metal strip that is formed into a tube and is filled continuously with powder. This process has consistently produced monofilamentary and multifilamentary wires of up to 4 km in length. This conductor can be supplied in a round or rectangular cross sections. MgB_2 is lighter weight and can be produced at a lower cost than the high temperature ceramic superconducting tape conductors (1G or 2G) when operated in the 20 K range. MgB_2 wire is versatile in that it can be sized (amperage and J_c) for the appropriate coil size and performance.

A cross section of a typical Hyper Tech wire and its performance in magnetic field is shown in Figure 2.16. This wire is manufactured 1 to

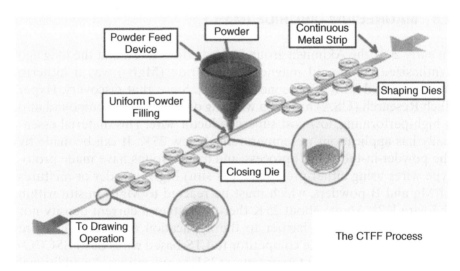

Figure 2.15 Hyper Tech MgB2 wire manufacturing process

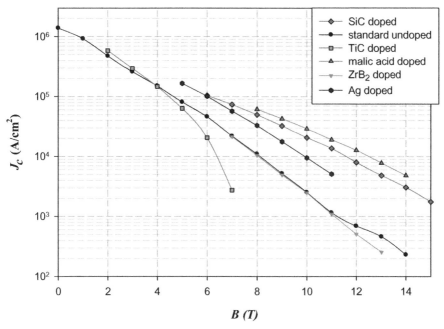

(a) 4.2 K high field (12 T) transport current

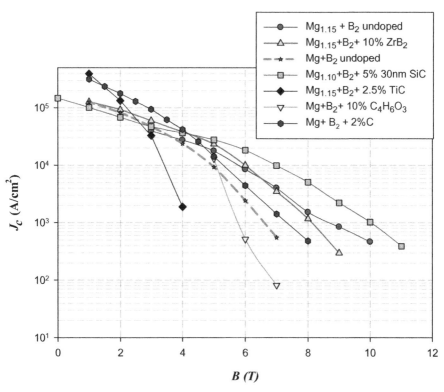

(b) 20 K high field (6 T) transport current

Figure 2.16 I_c properties of MgB$_2$ wire (Courtesy of Hyper Tech)

Table 2.5 MgB2 wire from Hyper Tech Research

Specification	
Critical current density at 20 K and 2 T	175 kA/cm²
Superconductor fraction	13–18% (in future 30%)
Diameter, mm	0.7–0.9
Number of filaments	7 and 19
Condition	Reacted and unreacted
Heat treatment temperature/time	700°C for 20 min (nominal)
Maximum allowable tensile strain at 77 K	0.35%
Continuous piece length, km	1–4

Source: Data from Hyper Tech Research website www.hypertechresearch.com in 2010.

4 km in length with 7 and 19 filaments. More details are provided in Table 2.5. In 2010 the performance of MgB_2 wires was still improving, and the reader is advised to contact wire manufacturers for the current wire specifications and capability curves.

This wire is very versatile and could have many potential applications in devices that can operate at temperatures below 25 K. Some of possible applications include transformers, inductors, reactors, motors, generators, fault current limiters, and magnetic resonance imaging (MRI) systems.

2.7 STATE-OF-THE-ART OF VARIOUS HTS CONDUCTORS

Currently three HTS materials (BSCCO-2223, YBCO-123, and MgB_2) are commercially available. However, their J_e as a function of magnetic field and operating temperature vary widely. Suitability of an HTS material for a given application should be evaluated using data similar to that shown in Figure 2.8. The development of the three materials in decreasing order of maturity was BSCCO-2223, YBCO-123, and MgB_2. BSCCO-2223 is the most developed but mechanically weak in its virgin condition. It must be laminated to a stronger material (e.g., stainless steel, brass or copper) to withstand mechanical forces experienced during the winding operation and in service. Another constraint of this material is its limited maximum length due to its manufacturing process. Then again, YBCO-123 coated conductors contain strong material (Hastealloy, nickel-tungsten, etc.) and could be available in longer lengths necessary for manufacturing commercially viable products. Both BSCCO and YBCO materials have homogeneity problems; that is, J_e along the length of a given conductor is not uniform. A section with the lowest J_e determines the operating current in a coil. Moreover J_e is a strong function of the orientation of the magnetic field with respect to the wide face of highly aspected conductors. J_e is higher when the magnetic field is llel to the wider face of a conductor, and lower when the magnetic field is perpendicular to it. Thus, in the design of coils for a given application, it is essential that this effect be carefully accounted for. The MgB_2 conductor is the least developed at the present time. It has the most desirable characteristic in that it can be made in a round shape and therefore be independent of the magnetic field orientation. Round conductors are also much easier to wind. This material is only suitable for applications at temperatures lower than 20 K and has low J_e. Its availability in long lengths is limited.

2.8   SUPERCONDUCTING MAGNET DESIGN

Several high-field magnets were built in late 1990s using BSCCO-2223 wire [13,14]. An example of designing a superconducting magnet is presented in this section. BSCCO-2223 wire is used in this example as it has the most characteristic data available. This magnet is designed to the specification in Table 2.6.

The first step in magnet design is to calculate the coil size and ampere turns for generating the required magnetic field on the magnet axis. In this example the space between the warm bore and the cold magnet inside radius is occupied by a metallic room-temperature wall and thermal insulation. This space is chosen as 25 mm on the basis of past experience of designing thermal insulation systems. Past experience determines the inner diameter of the cold magnet coil to be 150 mm. The cold coil's outside diameter and axial length are selected iteratively to produce the required field on the magnet axis. For an air-core magnet, the magnetic field can be calculated with analytical as well as finite-element codes. For this example a finite-element code is used for

Table 2.6   Specification for an example magnet

Parameter	Value
Clear warm-bore diameter, mm	100
Field on axis—at midpoint, T	3
Ramp time—zero to rated field, s	2
Operating temperature, K	35
Cooling method	Conduction
Cooler type	Cryocooler

Table 2.7 Coil design details

Parameter	Value
Coil bore diameter, mm	150
Radial build of the coil winding, mm	50
Axial length of the coil, mm	240
Ampere turns required, kA-turns	385
AMSC BSCCO-2223 bare wire dimensions	
• Width, mm	4.4
• Thickness, mm	0.29
Height of a pancake coil (includes insulations and structural support), mm	8
Number of pancake coils	30
Turns/pancake coil	70
Thickness of an insulated turn, mm	0.7
Total number of turns in the magnet	2100
Current per turn, A	185
Selected operating temperature, K	35
Peak field experienced by the conductor	
• $B_{\parallel}$ (T)	3.1
• $B_{\perp}$ (T)	1.1
Select AMSC BSCCO-2223 wire with I_c (self-field, 77 K), A	145
I_c at design $B_{\parallel}$ at operating temperature from Figure 2.6, A	280
I_c at design $B_{\perp}$ at operating temperature from Figure 2.6, A	230
Mean-turn length, m	0.63
Total wire needed for the coil, m	~1350
Inductance of magnet coil, H	0.83
Voltage to charge magnet in required time($= 2$ s), V	51
Maximum discharge voltage allowed, V	500
Minimum discharge time from full field, s	0.31

Source: I_c obtained from Figure 2.6 is scaled by 145/155 to correspond to the selected conductor.

performing magnetic field calculations. After a couple iterations the coil's outside diameter, axial length, and ampere turns needed to generate the required field of 3 T were selected. These values are included in Table 2.7. The coil design is carried out with AMSC BSCCO-2223 wire rated at 145 A (self-field, 77 K). This coil design consists of a set of pancake coils. The height of each pancake coil is determined as follows:

$$h_{pc} = \text{wire width} + \text{electrical insulation} + \text{axial support}$$
$$= 4.4 + 0.25 + 2.35 = 8 \text{ mm.} \tag{2.2}$$

Support is needed to bear axial stress developed in the coil during operation.

The radial thickness of each turn is calculated as

$$t_{turn} = \text{wire thickness} + \text{electrical insulation} + \text{hoop support}$$
$$= 0.29 + 0.25 + 0.16 = 0.7 \text{ mm.} \tag{2.3}$$

Since the coil radial build is determined to be 50 mm, it is possible to accommodate $50/0.7$ ($= t_{turn}$) ~ 70 turns in each pancake. The number of pancakes is calculated by coil length/pancake coil thickness = 240/8 ($= h_{pc}$) ~ 30. Thus the total number of turns in this coil are number of pancakes × turns/pancake coil = $30 \times 70 = 2100$ turns. The total number of ampere turns required in this coil is 385 kA. From this the current per turn is calculated as 385 kA/2100 turns ~ 185 A/turn.

The next step is to determine critical current of the selected conductor in presence of $B_{\parallel}$ and $B_{\perp}$ fields at the selected operating temperature. The finite-element field plot in Figure 2.17 shows that the peak $B_{\parallel}$

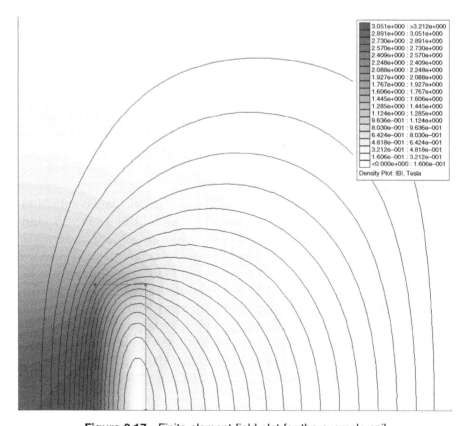

Figure 2.17   Finite-element field plot for the example coil

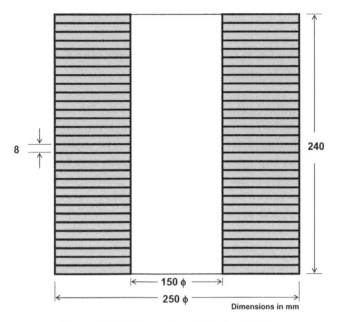

Figure 2.18   Example coil design details

and $B_\perp$ fields experienced by the coils are 3.1 and 1.1 T, respectively. From Figure 2.7 the I_c at these fields and the selected operating temperature of 35 K are 300 A and 250 A, respectively. Since the I_c values in Figure 2.7 are for a 155 A wire but the 145 A wire is selected for this coil, the I_c values to be used here should be scaled by factor 145/155. So the new values are 280 A and 230 A. The final coil design details are shown in Figure 2.18.

The third step in the magnet design process is to calculate conduction losses using equation (2.1). The finite-element field plot gives $B_\parallel$ and $B_\perp$ throughout the winding pack. It is therefore possible to determine the electric field $E(I)$ using the local $B_\parallel$ and $B_\perp$ magnetic field components. The I_c value to be used in equation (2.1) should be the lower of the two values calculated using local $B_\parallel$ and $B_\perp$. The total voltage V_e for a local element is the $E(I)$ calculated with local field values times the element length. The total coil voltage (V_t) is obtained by summing V_e values for all elements. The coil conduction loss (P_c) is equal to $V_t \times I$, where I is the coil current. It should be noted that $V_e \times I$ is the loss in a local element, which may be useful for assessing adequacy of cooling in a local region.

For this example magnet, let us assume that average $B_\parallel$ and $B_\perp$ are 2.5 and 0.75 T, respectively. The corresponding I_c values from Figure 2.6

for $B_{\parallel}$ and $B_{\perp}$ are 300A and 280A, respectively—these values are already scaled by the factor 145/155. The N-value for such a coil is expected to be around 16. Thus the $E(I)$ value from equation (2.1) is

$$E_e = 10^{-6} \times \text{V/cm} \times \left(\frac{I = 185\text{A}}{I_c = 230}\right)^{N=16} = 3.1 \times 10^{-8} \text{ V/cm}. \qquad (2.4)$$

Since the magnet has 1350 m of wire in it, the total loss in the coil is given as

$$P_c = E_e \times I \times \text{wire length}$$
$$= 3.1 \times 10^{-8} \text{ V/cm} \times 185\text{A} \times 1350 \text{ m} \times 100 \text{ cm/m} = 0.8 \text{ W}. \quad (2.5)$$

Thus the total conduction loss in this magnet is 0.8 W, which is quite low.

The inductance is 0.83 H for this example magnet, and the energy stored in it is 13.9 kJ when fully charged. To fully charge the magnet in 2 seconds (as required in Table 2.6), a DC voltage (V_{dc}) must be applied to the magnet:

$$V_{dc} = L \, dI/dt = \frac{0.83 \text{ H} \times 185\text{A}}{2 \text{ s}} = 77 \text{ V}. \qquad (2.6)$$

In the event of an emergency dump, 500V DC could be imposed on this system. The time (t_d) needed to discharge this magnet is calculated as

$$t_d = \frac{0.83 \text{ H} \times 185\text{A}}{500 \text{ V}} = 0.31 \text{ s}. \qquad (2.7)$$

The magnet coil assembly is subjected to hoop and axial forces while carrying the current. Total hoop and axial forces in this magnet are 116 and 110 kN, respectively. These forces are used for calculating stresses in the conductor as follows:

Hoop stress   Each HTS wire has 0.16 mm of external steel support (equation 2.3) plus 0.05 mm of steel in the wire as produced by AMSC. Thus total steel thickness in the wire is 0.16 + 0.05 = 0.21 mm, and each wire is 4.4 mm wide. Assuming that the hoop stress is uniformly distributed throughout the coil's cross section, the average hoop stress (ε_h) is calculated as

$$\varepsilon_h = \text{hoop force}/(\text{number of turns in the coil} \times \text{steel thickness}$$
$$\times \text{wire width})$$
$$= \frac{116\,\text{kN}}{2100 \times 0.21\,\text{mm} \times 4.4\,\text{mm}}$$
$$= 60\,\text{N}/\text{mm}^2 = 60\,\text{MPa}. \tag{2.8}$$

The HTS wire will also experience this stress, from Table 2.1, this is well below the maximum allowable tensile stress in the AMSC 1G wire.

Axial stress   The maximum compressive force will be borne by the pancake at the middle of the coil assembly. The average compressive stress in the HTS wire steel is calculated as

$$\varepsilon_c = \text{axial force}/(\text{number of turns} \times \text{mean-turn length}$$
$$\times \text{steel thickness})$$
$$= \frac{110\,\text{kN}}{2100 \times 0.63\,\text{m} \times 100\,\text{mm}/\text{m} \times 0.21\,\text{mm}}$$
$$= 4\,\text{N}/\text{mm}^2 = 4\,\text{MPa}. \tag{2.9}$$

This compressive stress in the steel and the HTS conductor is well within allowable limit.

Although the average tensile and compressive stresses calculated above are well within HTS conductor capabilities, local stresses could be higher because of the non-uniformity of the coil cross section, inclusion of nonmetallic components, such as electrical insulation and epoxy, that have much lower modulus of elasticity than steel. Local stresses and stress risers are critical design issues in all superconducting magnet.

2.9 SUMMARY

This chapter has presented information on the state of the art of high temperature superconductor technology. The chapter covered the most popular HTS conductor configurations, provided characteristic data that can be used for designing superconducting magnets, and further provided an example magnet design to help those readers who are new to this field. There are number of other important issues, such as hysteretic and eddy-current losses, and cryostat heat load, quench protection, and current lead performance and losses, that will be discussed in the later chapters that deal with specific device designs.

REFERENCES

1. Peter J. Lee, ed., *Engineering Superconductivity*, Wiley-Interscience, New York, 2001.
2. R. D. Blaugher, "Superconducting Electric Power Applications," *Adv. Cryogenic Eng.* 42: 883–898, 1996.
3. R. M. Scanlan, A. P. Malozemoff, and D. C. Larbalestier, "Superconducting Materials for Large Scale Applications," *Proc. IEEE* 92(10): 1639–1654, 2004. DOI 10.1109/JPROC.2004.833673.
4. E. Barzi, L. Del Frate, D. Turrioni, R. Johnson, and M. Kuchnir, "High Temperature Superconductors for High Field Superconducting Magnets," *Adv. Cryogenic Eng.* 52: 416, 2006.
5. K. R. Marken, W. Dai, L. Cowey, S. Ting, and S. Hong "Progress in BSCCO-2212/Silver Composite Tape Conductors," *IEEE Trans. Appl. Superconductivity* 7(2, Pt. 2): 2211–2214, 1997. DOI 10.1109/77.621033.
6. K. R. Marken, H. Miao, M. Meinesz, B. Czabaj, and S. Hong, "BSCCO-2212 Conductor Development at Oxford Superconducting Technology," *IEEE Trans. Appl. Superconductivity* 13 (2, Pt. 3): 3335–3338, 2003. DOI 10.1109/TASC.2003.812307.
7. L. J. Masur, J. Kellers, F. Li, S. Fleshler, and E. R. Podtburg, "Industrial High Temperature Superconductors: Perspectives and Milestones," *IEEE Trans. Appl. Superconductivity* 12(1): 1145–1150, 2002.
8. A. P. Malozemoff, S. Annavarapu, L. Fitzemeier, Q. Li, V. Prunier, M. Rupich, C. Thieme, W. Zhang, A. Goyal, M. Paranthaman, and D. F. Lee, "Low-Cost YBCO Coated Conductor Technology," *Superconductor Science & Technology* Vol. 13, No. 5, 473, 2000.
9. D. Verebelyi, E. Harley, J. Scudiere, A. Otto, U. Schoop, C. Thieme, M. Rupich, and A. Malozemoff, "Practical Neutral Axis Conductor Geometries for Coated Conductor Composite Wire," *Supercond. Sci. Technolo.* 16: 1158–1161, 2003. Ibid, "Uniform Performance of Continuously Processed MOD-YBCO Coated Conductors Using a Textured Ni-W Substrate," *Supercond. Sci. Technol.* 16: L19–L23, 2003.
10. X. Xiong, K. P. Lenseth, J. L. Reeves, Y. Qiao, R. M. Schmidt, Y. Chen, Y. Li, Y-Y Xie, and V. Selvamanickam, "High Throughput Processing of Long-Length IBAD MgO and Epi-Buffer Templates at SuperPower," Paper 4MA3. Presented at the IEEE Applied Superconductivity Conference, August 2006.

11. J. Nagamatsu, N. Nakagawa, T. Murakana, Y. Zenitani, and J. Akimitsu, "Superconductivity at 39 K in Magnesium Diboride," *Nature* 410: 63–64, 2001.

12. R. Flukiger, H. L. Suo, N. Musolino, C. Beneduce, P. Toulemonde, and P. Lezza, "Superconducting properties of MgB_2 tapes and wires," *Physica C* 385: 286–305, 2003.

13. G. Snitchler, S. S. Kalsi, M. Manlief, R. E. Schwall, A. Sidi-Yekhief, S. Ige, R. Medeiros, T. L. Francavilla, and D. U. Gubser "High-Field Warm-Bore HTS Conduction Cooled Magnet," *IEEE Trans. Appl. Superconductivity* 9(2, Pt. 1): 553–558, 1999. DOI 10.1109/77.783356.

14. H. Morita, M. Okada, K. Tanaka, J. Sato, H. Kitaguchi, H. Kumakura, K. Togano, K. Itoh, and H. Wada "10 T Conduction Cooled Bi-2212/Ag HTS Solenoid Magnet System," *IEEE Trans. Appl. Superconductivity* 11(1, Pt. 2): 2523–2526, 2001. DOI 10.1109/77.920379.

3

COOLING AND THERMAL INSULATION SYSTEMS

3.1 INTRODUCTION

All currently known superconductors must operate at cryogenic temperatures between 4 and 80 K. The technology of creating low temperatures is quite complex, and a great deal of ingenuity is required to build cooling equipment (called refrigerators or simply cryocoolers) capable of achieving these low temperatures. Discussion in this chapter is limited to the creation and maintenance of this cryogenic environment in which superconducting devices function. The discussion includes an introduction to gases suitable for the cryogenic environment, design of cryostats (or enclosures) for maintaining superconductors in their required cryogenic environment, and cryogenic refrigerators used for removing thermal load from the cold environment. Also discussed is the cryostat design process, including calculations of various heat loads. An example cryostat design is included at the end of this chapter for illustrating the design process and highlighting the key design drivers.

Applications of High Temperature Superconductors to Electric Power Equipment, by Swarn Singh Kalsi
Copyright © 2011 Institute of Electrical and Electronics Engineers

3.2 ANATOMY OF A CRYOSTAT

A cryostat houses and maintains superconductor magnet coils in an environment suitable for their operation. A cryostat assembly to house the HTS example magnet described in Chapter 2 is schematically shown in Figure 3.1. This HTS example magnet is epoxy impregnated and is conduction cooled with a cryocooler to maintain it at its required operating temperature of 30 K. The cryostat assembly consists of the following components; vacuum vessel, HTS magnet, current leads, cold mass support, intermediate temperature radiation shield, multi-layer insulation (MLI), and cryocooler. The vacuum vessel is configured to house the HTS magnet and to provide a clear 100-mm-diameter bore. The HTS magnet coil assembly (also called the "cold mass") is supported from the warm vacuum vessel. An intermediate temperature shield maintained at 60 K surrounds the magnet assembly. The purpose of this shield is to intercept radiated heat from the room-temperature wall of the vacuum vessel. A multi-layer insulation (MLI) thermal blanket is installed in the space between the radiation shield and room-temperature walls of the vacuum vessel. A pair of leads supplies current to the magnet. Since current leads represent a big source of heat conduction

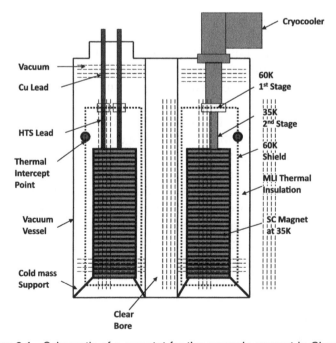

Figure 3.1 Schematic of a cryostat for the example magnet in Chapter 2

from room temperature to the cold mass, they are optimally designed to minimize this load. A typical current lead has two parts: a lower part (between the magnet and radiation shield) made from an HTS material, and an upper part (between the radiation shield and room temperature) made from copper or brass. The thermal load conducted from room temperature to the radiation shield is intercepted by the first stage of the cryocooler. Likewise the thermal load conducted from the radiation shield to the magnet is removed by the second stage of the cryocooler. Figure 3.1 shows a two-stage cryocooler employed for cooling the magnet assembly. The first-stage cooling intercepts the thermal load radiated from room temperature through the MLI thermal insulation, and conduction through the cold mass supports, through instrumentation wires, and through current leads at the selected 60 K radiation shield. Second-stage cooling removes thermal load in the 35 K region, which includes thermal radiation from the 60 K shield, and thermal conduction through cold mass supports, instrumentation wires, and conduction down HTS current leads between the first stage of the cryocooler and the magnet cold mass. Two-stage cryocoolers have even been used to cool a magnet to 4 K, which is necessary for LTS magnets. A two-stage cooler is selected for the example magnet to help illustrate various features of a cryostat assembly. However, a single-stage cryocooler is usually adequate to operate an HTS magnet down to about 30 K.

Figure 3.1 shows a cryogen free conduction-cooled magnet. It is conduction cooled with a cryocooler by providing a low thermal resistance path between the coil and the cryocooler. Cryocoolers cool down superconducting magnets from ambient temperature (300 K) to 30 K (or even 4 K for LTS coils.) In general, a cryocooler is employed for cooling superconducting magnets consisting of potted (epoxied) coils. Potted coils are conduction cooled even if they are bathed in a cryogen.

The design process for various cryostat components is described in the following sections.

3.3   CRYOGENIC FLUIDS FOR COOLING HTS MAGNETS

A superconducting magnet has to be cooled to a temperature suitable for its efficient and reliable operation. The simplest way of cooling a magnet is to bathe it in a pool of liquid cryogen, which boils at a temperature suitable for the superconductor materials. A cryogen boils at a constant temperature if it is maintained under a constant pressure.

Table 3.1   Selected properties of cryogens at 1 atmosphere

Property	He	H_2	Ne	N_2	O_2
Boiling temperature (T_s), K	4.22	20.39	27.09	77.39	90.18
Triple point, K	—	13.96	24.56	63.16	54.36
Heat of vaporization (liquid), kJ/kg	20.9	443	85.9	199.3	213
Heat of vaporization (liquid volume), J/cm³	2.6	31.1	104	161	243
Density (T_s, liquid), kg/m³	125	70.8	1206	807	1141
Density (T_s, vapor), kg/m³	16.9	1.33	9.37	4.60	4.47
Density (293 K), kg/m³	0.167	0.084	0.840	1.169	1.333
Density (T_s, liquid)/ Density (293 K)	749	843	1436	690	856

Source: Data taken from Iwasa [1].

The five fluids of interest (He, H_2, Ne, N_2, and O_2) are listed in Table 3.1) along with their key characteristic properties [1]. For gas properties, the reader is referred to the quoted reference in the table and many other available data sources. Of these fluids H_2 and O_2 are not preferred due to the fire hazard associated with them. Liquid O_2 can make materials (like Teflon or aluminum) explode and burn even though they are considered noncombustible. Hydrogen is used in limited high-tech applications where its safety can be carefully monitored. The remaining fluids (He, Ne, and N_2) are preferred because they are inert and do not combine with other fluids to create explosive mixtures.

Since most LTS magnets made with niobium-titanium (Nb-Ti) and niobium-tin (Nb_3Sn) superconductors require operation at close to 4 K, liquid He (LHe) has been the only fluid of choice. Almost all high-field LTS magnets employ LHe as a coolant. However, HTS conductors (BSCCO-2223 and YBCO-123) can operate at LN_2 temperature (77 K) for low-field applications and at LNe temperature (27 K) for high-field applications. LN_2 is preferred as a coolant because it is inexpensive, inert, and readily available. LNe is also attractive as a coolant for HTS magnets but is expensive and not readily available in large quantities. The main disadvantage of all cryogens is that they must be replenished or be used in a closed cycle where a cryocooler recondenses boiled-off vapor. Many HTS magnet applications, to be discussed in the later chapters, have used LNe and LN_2 as coolants. As a point of caution, all established safety and control rules must be followed while working with cryogens because of the hazards associated with them.

3.4   DIRECT COOLING WITH CRYOGENS

LTS magnets employing NbTi and Nb_3Sn superconductors are generally cooled by submerging them in LHe. Nearly all magnetic resonance imaging (MRI) magnets built with NbTi are cooled this way. Since the NbTi conductor is bathed in LHe, any heat generated in the conductor is transferred to the liquid coolant. This heat load is absorbed by converting LHe into gaseous He (GHe) using the latent heat of vaporization. In these magnets, superconductors remain in the superconducting state as long as the heat flux (heat per unit area) available at the conductor surface is less than what the LHe can remove; this limit is called the critical heat flux. Once the critical heat flux is exceeded, the superconductor temperature begins to rise, and eventually the superconductor transitions to its normal state when its temperature exceeds its critical temperature (T_c). The majority of MRI and particle accelerator magnets fall in this category.

Another class of magnets is cooled by a forced flow of liquid helium. In this method either superconductor strands are contained inside a conduit or they are attached to the outside of the conduit. LHe is forced through the conduit to cool the superconductor strands. This method permits more effective cooling of superconductor strands and in a more controlled manner than the pool boiling method. Very large magnets have been built using both approaches. For an in-depth study of behavior of LTS superconductors used as magnets, the reader is referred to a book by Wilson [2].

LTS superconductor wires are usually made in small diameter because of technical and manufacturing limitations. A typical Nb-Ti and Nb_3Sn wire (or strand) is about 0.5 to 1 mm in diameter. An individual wire may only carry about 100 A to 200 A, which in not suitable for making large magnets. A large magnet requires many ampere turns. If such a magnet were made using a small-diameter wire, then many turns would have to be wound. The resulting high inductance of such a magnet creates problems like quench protection, long charging time, and increased winding cost. A superconducting coil insulation could withstand only a limited voltage across its terminals due to conflicting electrical and thermal cooling requirements. Usually a maximum of about 1000 V is permitted across terminals of a coil. In order to protect a magnet, a shortest possible discharge time is desired to extract magnet energy following an abrupt quench. To achieve a lower inductance magnet, it is necessary to use high-current conductors, which are created by bundling many wires together—appropriately twisted to ensure

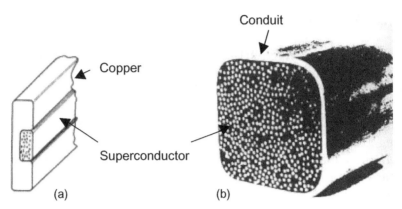

Figure 3.2   High-current superconductors: (a) Pool-boil cooling and (b) forced-flow cooling

equal current distribution among all wires. Figure 3.2 shows two such conductors:

1. cabled superconductor soldered into a copper conductor for use with pool-boiling cooling technique, and
2. cable-in-conduit conductor where many strands are housed in a conduit for use with forced-flow cooling technique.

Many variation of creating large current conductors have been successfully built. Cable-in-conduit type conductors are being considered by the ITER [3] project, which is a joint international research and development project that aims to demonstrate the scientific and technical feasibility of fusion power.

3.5   INDIRECT OR CONDUCTION COOLING

Many high-field magnets have been built using simple and convenient conduction cooling with cryocooler refrigerators. Such magnets exceeding 12T are commercially available now. Superconducting coils are epoxy impregnated to create monolithic structures that are mechanically strong and easy to handle. Such coils are cooled from their outside surface by conduction. Two typical conduction cooling schemes are shown in Figure 3.3, where good thermal contact is implemented between HTS coils and (1) tubes carrying coolant and (2) directly by a cryocooler refrigerator cooler.

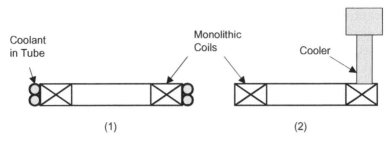

Figure 3.3 Conduction cooling of a superconducting coil: (1) Coolant in tubes and (2) coil in contact with a refrigerator cooler

The coolant in the cooling tubes could be a cold gas or a liquid cryogen. The coolant can operate in a closed cycle, rejecting its heat load outside the magnet system in a refrigerator or other equivalent cooler. The coolant could also be a cryogen taken from a storage container. Additionally a cryocooler refrigerator could be thermally interfaced to an HTS coil being cooled. The next section discusses various types of cooling devices currently available.

3.6 REFRIGERATION SYSTEMS

Refrigeration is the process of removing heat from an enclosed space, or from a substance, and rejecting it elsewhere for the primary purpose of lowering the temperature of the enclosed space or substance and then maintaining that lower temperature. Refrigerator systems require power to remove the thermal load from the low-temperature region. The *efficiency* (η) of cryogenic refrigerators has been improving steadily over many years. However, the efficiency of refrigerators of greatest interest in HTS is typically less than 20% of *Carnot efficiency*, with (η_c,) defined as

$$\eta_c = \frac{T_l}{T_h - T_l}. \tag{3.1}$$

For an ambient temperature ($T_h = 300\,\text{K}$) and an operating temperature ($T_l = 77\,\text{K}$), the η_c is 34.5%. The typical efficiency of a refrigerator is 20% of η_c, or $\eta = 6.9\%$. Another way to express this is through the *specific power* ($= 1/\eta$), which is 14.5 in this case. The specific power is expressed in units of watts per watt, namely the number of watts of input power required to remove one watt of heat at the cooling temperature. This means that to remove one watt of heat from the 77-K region, it requires 14.5 watts of input power. This requirement is also known as the *cryogenic penalty* because it is equivalent to a parasitic loss that diminishes the savings achieved with HTS conductors by eliminating the I^2R loss of a particular magnet or device. Obviously the η is lower at lower temperatures—for example, at 30 K, η_c is only 11%. Hence at lower operating temperatures, the actual operating efficiency is lower, the *specific power* of the refrigerator is higher, and the cryogenic penalty is more severe.

Radebaugh [4] has extensively reviewed the state-of-the-art of various refrigeration systems available for cooling magnets built with HTS and LTS conductors. Most magnets operate at temperatures in the range of 4 to 80 K. Refrigerators in this temperature range fall in the following two categories:

- *Recuperative types (steady flow).* Examples are the Joule–Thomson, Brayton, and Claude cycles.
- *Regenerative types (oscillating flow).* Examples are the Stirling, Gifford–McMahon [5], and pulse-tube cycles.

During the US DOE's 1000-hp motor development program with Reliance Electric, American Superconductor built a refrigerator using Reverse Brayton Cycle [6] for cooling the rotor. It generated a cold helium gas (around 24 K) that was used for the conduction cooling of the HTS coils in the motor. However, discussion in this section is limited to the regenerative-type coolers, which are widely employed for cooling HTS magnets. Specific refrigerators of interest are Gifford–McMahon, Stirling, and pulse tube.

3.6.1 Gifford–McMahon (G-M) Cryocoolers

The Gifford–McMahon (G-M) cryocooler is the most popular for cooling HTS magnets. One of its main advantages is it isolates the compressor from the regenerator and displacer, which allows a modified commercial air-conditioning compressor to be used. Maintenance is required about every one to three years. A G-M unit can be large and heavy, and there is still inherent vibration from the moving displacer. However, placing the compressor and the cold head remotely from the device being cooled can mitigate this problem.

G-M cryocoolers are widely used for cooling magnets from 2 to 80 K. These coolers are made as one-, two-, and three-stage devices and are made in various sizes in the United States, Europe, and Japan.

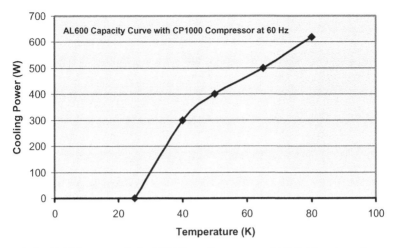

Figure 3.4 AL330 cryocooler's 100-W cooling power at 30 K (Courtesy Cryomech, Inc.)

Three-stage coolers are usually for generating temperatures below 2 K. For most magnet applications, one- and two-stage coolers suffice. Usually two-stage coolers are employed to achieve cooling down to 4 K, whereas single-stage coolers are sufficient for temperatures at 20 K and above. Thus HTS magnets usually employ single-stage coolers. A large capacity single-stage Cryomech AL600 cryocooler can supply cooling power [7] as shown in Figure 3.4 when coupled with their CP-1000 compressor. The specific data in Table 3.2 is for this cooler and its associated compressor (data from the website www.cryomech.com). A typical cryocooler and its associated compressor are shown in Figure 3.5.

Although G-M cryocoolers are popular and are available in very small to large capacity rating, they have following drawbacks:

· Efficiency
· Size and weight
· Noise and vibrations
· Heat rejection
· Cost

These issues warrant careful evaluation before selecting a G-M cooler for a given application. However, G-M coolers are quite portable and versatile for a large variety of applications and have made possible

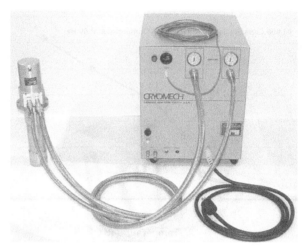

Figure 3.5 Cryocooler cold head and its compressor (Courtesy Cryomech, Inc.)

Table 3.2 Data for a large capacity cryocooler

Parameter	
Cold head model	AL600
Weight, kg	47.6
Dimensions—cold finger diameter × height, mm	159ϕ × 308
Water-cooled compressor	CP1014
Weight, kg	222.7
Dimensions—L × W × H, mm	610 × 580 × 970
Cooling type	Water

Source: Data from the Cryomech website, www.cryomech.com [7].

many HTS based devices. About 20,000 units per year were made worldwide during the peak in the semiconductor industry. The G-M cryocooler is also used for cooling radiation shields up to 10 to 15 K in MRI systems to reduce the boil-off rate of liquid helium or in some cases used to re-liquefy the helium at 4.2 K. Single-stage units for temperatures above 30 K are less expensive. Refrigeration powers at 80 K generally range from about 10 to 500 W, with input powers ranging from about 800 W to 10 kW. The oil-lubricated compressors have lifetimes of at least five years, but the absorber cartridge for oil removal must be replaced once every year or two. Replacement of seals on the displacer must be performed about once a year.

3.6.2   Stirling Coolers

Although the Stirling coolers have higher efficiency than G-Ms and can provide substantial cooling power at 65 K, they are still very heavy and cost substantially more. There is considerable manufacturing experience with such units, but they are prone to vibration, owing to the moving displacer. They must be run dry (without lubrication) because of the links (via the coolant gas) between cold and warm regions, and it is expensive to achieve long lifetimes (3–10 years) in these systems.

A Stirling cooler works by repeated heating and cooling of a sealed amount of working gas, usually helium for cryogenic temperatures. A piston varies the working gas volume, and a displacer shuttles the gas within the cooler between the warmer components and the cooler components. Stirling coolers are available in a wide range of sizes— from milli-watts, where they can be very small and relatively inexpensive, to hundreds of watts of cooling capacity. Temperatures down to 20 K are possible with two-stage units.

Stirling Cryogenics and Refrigeration makes three different types of coolers (described below) that are suitable for a range of requirements and temperature levels.

The *re-liquefaction Liquid Power Cooler (R-LPC)* cools down to 65 K and re-condenses the boil-off gas returning from the application. This power cooler type is the solution for applications where

- the design of the HTS device allows sufficient cooling by natural flow, and
- the boil-off gas developed in the application easily collects at the top of the application cryostat.

The *forced Flow Liquid Power Cooler (LPC)* cools down to 65 K using a subcooled liquid gas loop. This LPC type is advantageous for those applications where

- the refrigerator cannot be located in the immediate vicinity of the application, or
- the HTS device design does not allow sufficient cooling by sheer natural flow.

The *gas power cooler (GPC)* cools at 20 to 40 K and 80 K simultaneously, using a gaseous helium loop to transfer the cold to the

Table 3.3 Data for large capacity Stirling coolers

Parameter	SPC-1T	SPC-4T
Number of cylinders	1	4
Cooling capacity at the two stages	—	—
20 K stage, W	50	200
80 K stage, W	150	600
Cooling water requirement, l/hr	750	3000
Power input, kW	12	45
Installation space requirement, m^3	3	3
Total weight, kg	600	1800
Maximum ambient temperature, °C	45	45

Source: Data from Stirling Cryogenics and Refrigeration website, www.stirling.nl.

Figure 3.6 Large SPC-1T Stirling cooler (Courtesy Stirling Cryogenics and Refrigeration website)

application. The GPC is the solution for those applications that demand full refrigeration below 65 K.

Data for two low-temperature two-stage coolers is summarized in Table 3.3. A two-stage, one-cylinder cooler is shown in Figure 3.6.

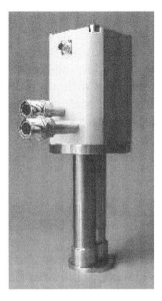

Figure 3.7 PT60 Cold Head (compressor package not shown)—Courtesy of Cryomech Inc.

3.6.3 Pulse-Tube Coolers

The advantage of a pulse-tube cooler is that the "displacer" is a column of gas, not solid material; it is a gas plug. This eliminates a crucial moving part at low temperatures, and greatly enhances reliability and reduces vibration. Most pulse-tube cryocoolers built to date have had small cooling capacities (50 W or less), but recent advances have demonstrated the feasibility of systems with up to 1 kW of cooling capacity at 77 K, and much larger capacities are expected in the future.

Current pulse-tube development efforts are concentrated on the basic understanding of the cooling principle, the extension of the temperature range, and the exploration of various cooling arrangements. Challenges are related to the increase of the coefficient of performance, the increase of the operating frequencies for low-temperature applications, the extension of the temperature range to lower temperatures, to long-term stability, and to miniaturization. These pulse-tube coolers are made in different cooling capacity ratings—from a few watts to tens of watts. Larger rating coolers are beginning to become available. A high-capacity commercial single-stage Cryomech pulse-tube cooler is shown in Figure 3.7. Its cooling power is 60 W at 80 K.

This PT60 Cryomech cryocooler has low vibrations and provides cooling to nearly 30 K. Vibration levels on the cold end have been measured as low as 1 μ when using the low-vibration option. The PT60 cooler is suitable for re-condensing liquid nitrogen and for cooling HTS devices. It could be supplied with a remote motor for minimizing the vibrations. These pulse-tube coolers require virtually no maintenance; merely change the compressor absorber every 20,000 hours. A PT60 cooler in single-stage configuration provides capability of 0 W at 30 K and 60 W at 80 K. Their compressors are cooled with air or liquid.

3.7   OPEN LOOP COOLING WITH LIQUID NITROGEN

There is a large liquid nitrogen production and supply infrastructure in place throughout the world with literally thousands of tons of available LN_2. This can be utilized in HTS applications with very attractive economics. Such systems could be simple to install and operate with very few complex components. Temperatures in the range of 65 to 77 K can be achieved with vacuum pumps.

There are many apparent advantages to an open loop system with LN_2 evaporating in *pool boiling*. The capital costs of such systems are low, and operational reliability is significantly higher because no major machinery is needed and it is a simple system. Still, while LN_2 costs are very competitive and the operating economics look superior to small mechanical refrigeration systems, costs can be higher in some remote locations. Also geometries of some systems may not favor forced circulation cooling over pool boiling and therefore may need special cryostat designs.

3.8   MAGNET MATERIALS

A magnet system consists of a variety of materials such as copper, brass, aluminum, steel, Inconel, Kevlar, Teflon, Kapton, Nylon, G-10 fiber glass, glass fabric, lead, Indium, and sapphire. Properties (electrical, mechanical, and thermal) are available from the website of National Institute of Standards and Technology (NIST), file:///C:/Work%20Files/Book/Chapter-3-Cooling%20and%20Thermal%20Ins/Chap_3-References/material_properties_NIST.htm. These properties are provided as fitted functions of temperature, which could be useful for integration in automated calculation routines. Iwasa [1] has included properties (electrical, mechanical, and thermal) of most commonly used materials in his book. Material properties are also available on the site http://www.submm.caltech.edu/cso/receivers/thermal/properties.html.

3.9 CURRENT LEADS

Current leads convey current from room temperature to cold magnet coils. These leads also form a thermal conduction path between warm and cold ends. Because of large cryogenic penalty to remove the heat load from the low temperature, it becomes mandatory to minimize the thermal conduction from high temperature to the low temperature. Leads can be divided into three groups (shown in Figure 3.8) according to cooling techniques employed:

1. *Conduction cooled* lead (Figure 3.8*a*) is cooled by removing thermal heat load from the cold end.
2. *Vapor cooled* lead (Figure 3.8*b*) is constructed by a employing porous metal structure through which cold gas vapor flows from the cold end to the warm end.
3. *Forced cooled* lead (Figure 3.8*c*) is also constructed by employing a porous metal structure through which gas is forced from the cold end to the warm end.

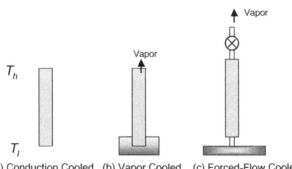

(a) Conduction Cooled   (b) Vapor Cooled   (c) Forced-Flow Cooled

Figure 3.8   Current leads cooled by (*a*) conduction, (*b*) vapor, and (*c*) forced flow

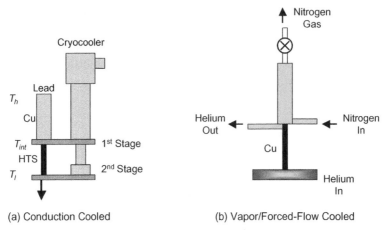

(a) Conduction Cooled   (b) Vapor/Forced-Flow Cooled

Figure 3.9   Hybrid current leads: (*a*) Conduction cooled and (*b*) vapor/forced-flow cooled

3.9.1   Current Leads for Low Temperature Magnets

To minimize heat conduction further, hybrid leads could be employed for magnets operating at very low temperatures. Two possible variations of such leads are shown in Figure 3.9.

1. Conduction cooled hybrid lead (Figure 3.9*a*) has two parts. The lower part operates between the magnet operating temperature ($T_l < 60\,\text{K}$) and an intermediate temperature ($T_{int} \sim 60\,\text{K}$). The upper portion operates between T_l and T_h (~300 K ambient) temperatures. Such leads are usually employed in cryogen-free magnets cooled with two-stage cryocoolers as shown in the figure. The lower portion is made from a suitable HTS material having a intrinsically low thermal conductivity and is conduction cooled by the second stage of the cryocooler. The upper portion is made of copper and is conduction cooled by the first stage of the cryocooler.

2. Vapor/forced-flow cooled lead (Figure 3.9*b*) also has two parts— each part could be vapor or force-flow cooled. The lower portion is cooled with vapor (helium) from the magnet. The vapor is allowed to warm up to the liquid nitrogen temperature. The upper portion is cooled with liquid nitrogen, which is allowed to warm up to ambient temperature ($T_h = 300\,\text{K}$). It is also possible to build the lower portion of the lead from HTS for minimizing the thermal load further.

Figure 3.10   HTS current leads (Courtesy of American Superconductor Corporation)

Most of the HTS magnets are built to operate dry and are conduction cooled with cryocooler. Such magnets cooled with a single cryocooler operate in 30 to 40 K temperature range employ conduction cooled leads (Figure 3.8a). However, magnets operating at lower temperatures (<20 K) employ two-stage cryocoolers and use leads like those shown in Figure 3.9a.

HTS leads are commercially available for a range of currents. There are two basic technologies for HTS leads—bulk rods of ceramic superconductor and metal matrix superconducting composites. There are advantages and disadvantages to each.

Bulk ceramic leads are made by a variety of methods and of a number of different HTS materials, but the primary objectives are the same: to achieve a rugged ceramic structure with high critical current and low electrical resistance connections. The advantage of this approach is that because the ceramics have intrinsically low thermal conductivity, the leads may be made quite short for easier integration in the system. The disadvantages are that the ceramic rods are susceptible to breakage during installation, during operation, and during temperature excursions caused by driving the lead normal. In addition it has proved difficult to provide very low resistance connections between the ceramic superconductor and the metallic connections at the ends of the leads. These disadvantages have been mitigated by careful system design in a number of magnets, and successful systems have been built using bulk leads supplied by Nexans Superconductor.

Metal matrix composite leads (Figure 3.10) essentially use the powder-in-tube technology of BSCCO-2223 wire to manufacture a wire or tape incorporating a low thermal conductivity metal or alloy in place of the customary silver matrix. The advantages of metallic leads are intrinsic ruggedness, high tolerance to thermal excursions, and very low contact resistances. These advantages are to be balanced against the disadvantage of the somewhat higher thermal conductivity of the composite material, which requires that a longer lead assembly be used to achieve heat leaks comparable to bulk leads.

3.9.2 Design of Conduction Cooled Leads

A conduction cooled lead (Figure 3.8*a* and upper portion of Figure 3.9*a*) is employed between 60 and 300 K (ambient temperature) regions. This portion is usually made from copper or brass and is designed to minimize thermal conduction from the warm end to the cold end. Simple equations governing this design are provided below. Total thermal conduction (Q) from the warm end to the cold end is given as

$$Q = \frac{\rho L I^2}{A} + \frac{K A \Delta T}{L} \tag{3.2}$$

where
  I = current in the lead,
  L = length of the lead,
  ρ = average resistivity of the lead material,
  K = average thermal conductivity of the lead material,
  A = area of cross section of the lead,
  ΔT = temperature difference between warm and cold ends.

Equation (3.2) can be rearranged as

$$Q = \left(\sqrt{\frac{\rho L I^2}{A}} - \sqrt{\frac{K A \Delta T}{L}} \right)^2 + 2I \sqrt{\rho K \Delta T} \tag{3.3}$$

To achieve minimum thermal conduction (Q), the first bracketed term in equation (3.3) must be made zero, which makes the thermal conduction equal to $I^2 R$ loss in the lead. With this assumption and upon rearrangement of terms, the following equation is achieved:

$$\frac{L}{A} = \frac{1}{I}\sqrt{\frac{K}{\rho}}\Delta T \qquad (3.4)$$

Equation (3.4) gives length/cross-sectional (L/A) ratio of a conduction cooled current lead as a function of lead current, temperature difference between warn and cold ends, resistivity, and thermal conductivity of the lead material. The thermal conduction (Q) to the lower end of a lead designed with equation (3.4) is given as

$$Q = 2I\sqrt{\rho K \Delta T} \qquad (3.5)$$

The product ρK is almost identical for a variety of material used for manufacturing current leads. Thus selection of lead material is less critical. Typical conduction heat load for operation between 60 and 300 K is about 4 W for a 100 A lead. For more accurate design analysis, using material properties as a function of temperature, readers are referred to the book by Iwasa [1].

3.10   EXAMPLE CRYOSTAT DESIGN

The design discussion in the previous sections is utilized here for sizing the cryostat and refrigeration system for the magnet example of Chapter 2, Section 2.7. The magnet and its cryostat assembly are shown in Figure 3.1. Design calculations are performed for calculating thermal load from various sources, and then this thermal load is used for selecting a suitable refrigerator.

3.10.1   Configuration

The magnet assembly has a clear warm bore of 100 mm. The HTS magnet is supported from the bottom of the vacuum vessel through a speaker cone-shaped structure that has many holes in its wall for minimizing the cross section for thermal conduction from the ambient temperature vacuum vessel to the cold mass. A two-stage current lead is included to supply current from ambient to the cold magnet. A radiation shield is included in the space between the warm walls of cryostat and the cold surface of the magnet. This radiation shield is operated at 60 K, which is also the temperature of intermediate stage of the current lead.

3.10.2 Thermal Load Calculations

Total thermal load of a conduction cooled cryogen free magnet has the following four components:

1. Conduction loss in the HTS wire (in the HTS coils) is 0.8W (already calculated in Section 2.8, equation 2.5).
2. Radiated thermal load through MLI.
3. Thermal conduction through current leads.
4. Thermal conduction through mechanical support components.

Radiated Thermal Load through MLI Thermal load radiated from ambient temperature (300K) vacuum wall to the radiation shield at 60K is calculated in this section. The four components of this thermal load are calculated below:

1. *Thermal load from the ambient cryostat walls to the radiation shield in the bore.* The HTS coil inside diameter is 150mm, and the height of the magnet is 240mm. The radial gap between warm cryostat wall and magnet cold wall is

$$L_{gap} = 0.5 \cdot (150 - 100)\text{mm} = 25 \text{ mm}. \qquad (3.6)$$

For a radiation shield that is located midway in the 25-mm gap, the diameter of the radiation shield is

$$d_s = 100 + 25 = 125 \text{ mm}. \qquad (3.7)$$

The height of the warm bore tube, h_b = magnet height (240mm) + gap between bottom of cryostat and the magnet (75mm) + gap between the top of the magnet and the top plate of cryostat (250mm) = 565mm.
The mean surface area of the MLI blanket is

$$A_{b\text{-}MLJ} = \pi \cdot d_s \cdot h_b = \pi \cdot 125 \text{ mm} \cdot 565 \text{ mm}$$
$$= 221762 \text{ mm}^2 = 0.222 \text{ m}^2. \qquad (3.8)$$

Typically thermal flux due to radiation from 300K warm wall to the 60K radiation shield (through MLI blanket) is about

$$Q_r = 3 \text{ W/m}^2. \qquad (3.9)$$

This flux can vary with number of MLI layers, the packing density, and the vacuum level. Typically a packing density of approximately 1 layer/mm and a vacuum level of 10^{-4} Torr or less are required for proper MLI performance. Therefore the total thermal load appearing at the radiation shield is

$$Q_b = Q_r \cdot A_{b\text{-}MLI} = 3 \cdot 0.222 = 0.67 \text{ W} \qquad (3.10)$$

2. *Thermal radiation from cryostat outer walls to the outer radiation shield.* The diameter of the cryostat's outer wall is

$$D_{o\text{-}CS} = \text{magnet OD } (250) + 2 \cdot 75 = 400 \text{ mm.} \qquad (3.11)$$

The diameter of radiation shield between magnet and the outer wall of the cryostat is

$$D_{o\text{-}RS} = D_{o\text{-}CS} \text{-}50 \text{ mm} = 400 - 50 = 350 \text{ mm.} \qquad (3.12)$$

The surface area of outer radiation shield is

$$A_{o\text{-}MLI} = \pi \cdot D_{o\text{-}RS} \cdot h_b = \pi \cdot 350 \cdot 565 = 0.621 \text{ m}^2. \qquad (3.13)$$

The thermal radiation through the MLI blanket in the outside radiation shield is

$$Q_o = A_{o\text{-}MLI} \cdot Q_r = 1.86 \text{ W.} \qquad (3.14)$$

3. *Thermal radiation from cryostat bottom to the magnet.* The surface area of the radiation shield is

$$A_{bot} = \frac{\pi}{4} \cdot (D_{o\text{-}RS}^2 - d_s^2) = \frac{\pi}{4} \cdot (350^2 - 125^2) = 0.084 \text{ m}^2. \qquad (3.15)$$

The thermal load due to radiation from the bottom of cryostat is

$$Q_{bot} = A_{bot} \cdot Q_r = 0.084 \text{ m}^2 \cdot 3 \text{ W/m}^2 = 0.25 \text{ W.} \qquad (3.16)$$

4. *Thermal radiation from cryostat top flange to the magnet.* To the first order, thermal radiation from the top flange of cryostat to the magnet can be assumed to be equal to the thermal load from the bottom:

$$Q_{top} = Q_{bot} \qquad (3.17)$$

The total thermal load due to radiation is given as

$$Q_{rad} = Q_b + Q_o + Q_{bot} + Q_{top}$$
$$= 0.67 + 1.86 + 0.25 + 0.25 = 3.87 \text{ W}. \tag{3.18}$$

Thermal radiation from 60 to 35 K surfaces should be negligible for a well-designed shield. However, substantial losses can occur in areas where holes in the shield are required for leads, instrumentation, supports, and the cryocooler. Care needs to be taken in the shield design to mitigate these potential loss areas.

3.10.3 Current Leads

The magnet is supplied with a pair of leads, each carrying 185 A. Brass is select as the lead material that has an average resistivity (ρ) of 0.04 $\mu\Omega$-m and a thermal conductivity (K) of 100 W/m/K. Assuming that the lower end of the lead is anchored at the radiation shield temperature of 60 K, the temperature difference between the ambient end (300 K) and shield end (60 K) is $\Delta T = 240$ K. From equation (3.4), the length/cross-sectional area (L/A) ratio for an optimum lead is

$$\frac{L}{A} = \frac{(K \cdot \Delta T / \rho)^{0.5}}{I}$$
$$= \frac{(100 \cdot 240 / (0.04 \cdot 10^{-6}))^{0.5}}{185} = 4187 \text{ 1/m}. \tag{3.19}$$

The length (L) of each lead (between 300 and 60 K ends) is about 0.4 m. Thus the cross-section (A) of each lead is

$$A = \frac{0.4}{4187} \text{ m}^2 = 95.5 \text{ mm}^2, \tag{3.20}$$

which translates to a lead diameter of 11 mm.

Thermal conducted through both leads to the 60-K end is calculated with equation (3.5):

$$Q_{lead} = 2 \cdot 2 \cdot I \cdot (\rho \cdot K \cdot \Delta T)^{0.5}$$
$$= 2 \cdot 2 \cdot 185 \cdot (0.04 \cdot 10^{-6} \cdot 100 \cdot 240)^{0.5} = 23 \text{ W}. \tag{3.21}$$

The lead section between the 60 and 35 K magnet employs HTS and is supported with a material that has low thermal conductivity. Thermal conduction from 60 to 35 K can be neglected.

3.10.4 Conduction

The magnet is supported from the bottom of the vacuum vessel using a speaker-type cone support that has many holes in its wall for reducing thermal conduction. This support is made from G-10 fiberglass material. The support cone has an average diameter of $d_s = 250\,$mm, thickness $t_s = 3\,$mm and height $h_s = 75\,$mm. The average thermal conductivity of G-10 is $K_s = 0.6\,$W/m/K. Total thermal conduction from ambient (300 K) to 60 K shield is

$$Q_{300\text{K-s}} = K_s \cdot (\pi \cdot d_s \cdot t_s) \cdot \Delta T / h_s$$
$$= 0.6 \cdot \left(\pi \cdot 0.25 \cdot \frac{3}{1000} \right) \cdot (300 - 60) / \left(\frac{75}{1000} \right) = 4.5 \text{ W}, \qquad (3.22)$$

and thermal load conducted to 35-K magnet mass is

$$Q_{\text{s-35K}} = K_s \cdot (\pi \cdot d_s \cdot t_s) \cdot \Delta T / h_s$$
$$= 0.6 \cdot \left(\pi \cdot 0.25 \frac{3}{1000} \right) \cdot (60 - 35) / \left(\frac{70}{1000} \right) = 0.5 \text{ W}. \qquad (3.23)$$

3.10.5 Selection of Refrigerator

Total thermal loads in 60 and 35 K regions are summarized in Table 3.4. These loads are used for selecting a suitable cryocooler.

A two-stage cryocooler with cooling capacity of 2 W at 35 K and 50 W at 60 K could be selected for this magnet. However, a two-stage cryocooler cooler is more expensive and less reliable than a single stage cryocooler. It is possible to eliminate the radiation shield and remove

Table 3.4 Summary of thermal loads

Thermal Load Component	35 K Region	60 K Region
Conduction loss in HTS, W	0.8	
Current leads, W	—	23
Radiation, W	—	3.87
Thermal conduction, W	0.5	4.5
Total, W	1.3	31.4

all losses from the 35 K region. The total thermal load in the 35 K region would be around 35 W. This thermal load could be handled with a Cryomech cooler AL230 that provides 60-W cooling at 30 K and is powered with a CP950 compressor, which consumes 5.5 kW of electric power.

3.11 SUMMARY

This chapter has provided a very basic introduction to cryostat design, reviewed available coolers, and presented an example of designing a cryostat and its cooling system for the magnet discussed in Chapter 2. The performance of the cooling system mainly determines the acceptance of the HTS devices in industry. Some of the other important factors relating to the cooling system are cost, serviceability, maintenance, compactness, and high refrigerator efficiency. More details are available in the references quoted at the end of this chapter and from manufacturer's websites.

REFERENCES

1. Y. Iwasa, *Case Studies in Superconducting Magnets: Design and Operational Issues*, Plenum Press, New York, 1994.
2. M. N. Wilson, *Superconducting Magnets*, Clarendon Press, Oxford, 1983.
3. M. Huguet, "Team and Home Teams, Key Engineering Features of the ITER-FEAT Magnet System and Implications for the R&D Programme," 18th IAEA Fusion Energy Conference, IAEA-CN-77/OV6/1, 2000.
4. R. Radebaugh, "Refrigeration for Superconductors," *Proc. IEEE* 92(10): 1719–1734, 2004.
5. W. E. Gifford, "The Gifford-McMahon Cycle," 1965 Cryogenic Engineering Conference, Rice University, Houston, August 23–25, 1965.
6. D. Aized, B. B. Gamble, A. Sidi-Yekhlef, J. P. Voccio, D. I. Driscoll, B. A. Shoykhet, and B. X. Zhang, "Status of the 1000 HP HTS motor development," *IEEE Trans Appl. Superconductivity* 9(2, Pt. 1): 1197–1200, 1999. DOI 10.1109/77.783514
7. Data from Cryomech website on January 31, 2010.

4

ROTATING AC MACHINES

4.1 INTRODUCTION

According to the US Department of Energy, motors account for 70% of all energy consumed by the domestic manufacturing sector and use over 55% of the total electric energy generated in America. Large electric motors, those greater than 1000 horsepower, consume over 30% of the total generated electric energy and 70% of these motors are suited to utilize high temperature superconductor (HTS) technology. With some minor exceptions, nearly all cruise ships today are being built with electrical propulsion, and many other types of commercial vessels and warships are adopting marine motors as their primary source of motive power.

Rotating machines (motors and generators) employ copper windings on rotor and stator. Current in these windings create resistive losses, which represent significant waste of energy and economic resources. Thus any efficiency improvement is likely to yield huge savings in energy use. Superconductivity offers zero to near-zero resistance to electrical current flow; thus the use of superconducting materials significantly reduces electrical energy loss as well as producing a reduction in size and weight of power components and machinery. The discovery

Applications of High Temperature Superconductors to Electric Power Equipment, by Swarn Singh Kalsi
Copyright © 2011 Institute of Electrical and Electronics Engineers

of HTSs in 1986 has provided the impetus to develop a variety of rotating machines employing superconducting windings. These HTS conductors operate at higher temperatures (between 25 K and 77 K) that simplify the refrigeration-cooling systems. Recently full-scale motors and generators have been demonstrated. This chapter describes topologies of ac rotating machines and their theory, design and manufacturing issues. Both high-speed and low-speed motors and generators are discussed. Finally, machine analysis is presented to enable simulations of these machines in an electric grid using industry standard computer codes.

HTS motors are ideal for use in pumps, fans, compressors, blowers, and belt drives deployed by utility and industrial customers, particularly those requiring continuous operation. They are suitable for large process industries such as steel milling, pulp and paper processing, chemical, oil and gas refining, mining and other heavy-duty applications. An important and rapidly growing use for HTS motors is in transportation applications, particularly naval and commercial ship propulsion, where size and weight savings provide a key benefit by increasing design flexibility and opening up limited space for other uses.

4.2 TOPOLOGY

Two popular types of electric rotating machines are synchronous and induction. A synchronous rotating machine has two windings: an AC winding located in the stator and a DC winding located on the rotor. This is the most common configuration, although the two winding locations are interchangeable. An induction motor has a squirrel cage or three-phase wound winding on the rotor. The rotor winding carries AC at slip frequency; the rotor frequency is equal to the line frequency when the slip is equal to 1 (rotor stationary) and the slip frequency (typically <5% of line frequency) when the motor is operating at its nominal speed. Induction motors are very popular in industry for lower ratings (<500 hp), but synchronous motors are preferred in larger sizes both in industry and on ships. Moreover induction motor windings, both on rotor and stator, experience AC currents and are therefore not good candidates for superconductor windings. As discussed in Chapter 2, superconductor losses are negligible only when the motor carries DC. These losses in the AC environment are quite large and are difficult to remove economically. Because of current high cost of superconductors and cooling systems, the primary applications of these motors are in large sizes (>1000 hp) for pump and fan drives

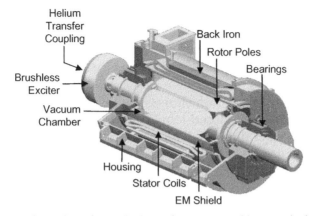

Figure 4.1 Configuration of a typical synchronous machine employing superconducting field winding on the rotor (Courtesy of American Superconductor Corporation)

for utility and industrial markets, particularly applications requiring
continuous operation, and large ship drives.

Figure 4.1 shows a typical AC synchronous machine with superconducting field winding. The rotating HTS field winding creates a magnetic field in the copper armature winding. The magnitude of this field
is typically twice that of a conventional motor. The HTS motor has an
air-core (i.e., nonmagnetic) construction on rotor and non-metallic
teeth in the stator, which enables the air-gap field to be increased
without the core loss and saturation problems inherent in laminated
iron stator and rotor cores. The copper armature winding lies just
outside the air gap—in some applications it is embedded in nonmetallic
teeth to provide mechanical support.

Under steady-state operation, the rotor spins in sync with the rotating field created by the three-phase armature currents, and the superconducting field winding experiences only DC magnetic fields. Under
load or source transients, however, the rotor moves with respect to the
armature-created fields, and it experiences AC field harmonics. An
electromagnetic (EM) shield located between the HTS coils and the
stator winding shields the HTS field winding from these AC fields. A
warm (room temperature) EM shield is positioned at the outer most
surface of the rotor. Inside the warm shield is a thermal insulation space
(vacuum) that surrounds the rotor cryostat. The cold EM shield (if
employed) is on the inside surface of this vacuum space and is a high-
conductivity shell near the operating temperature of the superconducting coils. The superconducting field coils are located within the inner
EM shield on a nonmagnetic support structure. The warm EM shield,

which directly transfers torque to the warm shaft, is designed to be mechanically robust to withstand the large forces generated during faults, and is designed to absorb heating caused by negative sequence currents and any other harmonic currents generated by a variable speed drive (VSD).

A refrigeration system, which uses cold circulating helium gas (or other suitable gas) in a closed loop, maintains the HTS field winding at cryogenic temperature. Helium gas is circulated through cooling channels located inside the rotor. The closed cooling loop runs from the turning rotor body to externally located, stationary refrigerator system utilizing Gifford-McMahon (G-M) cold heads discussed in Chapter 3.

The copper stator winding employs no magnetic iron teeth and is typically designed with class F (155°C maximum operation) insulation but is operated at class B (130°C maximum operation) insulation temperatures. To reduce eddy-current losses, the stator coils employ copper Litz conductor, which is made up of small-diameter insulated and transposed wire strands. Because the stator winding bore surface experiences a high magnetic field that would saturate the iron teeth of a conventional stator, the superconductor motor's stator armature winding does not employ iron teeth. With no iron teeth in the winding region, the support and cooling of the stator coils require special attention. The back EMF in air-gap stator winding is nearly a pure sine wave, and the harmonic field components are much smaller than those observed in the conventional machines.

4.3 ANALYSIS AND PARAMETER CALCULATIONS

An optimal design of a superconducting machine is based on judicial selection among competing and conflicting electrical, mechanical, thermal, economic, and reliability requirements. For example, thermal performance improves as the operating temperature is increased at the expense of reduced electrical performance. Electrical system stability and power density increases at the expense of structural stability. Thermal isolation improves at the expense of mechanical strength and vibration tolerance. The increased complexity of the superconductor technology must not compromise reliability. Finally, economic feasibility trumps all other compromises. Discussion in this section is limited to the description of basic theory of AC synchronous machines. Design equations developed here are utilized for sizing a synchronous machine with superconductor field winding. No distinction is made between

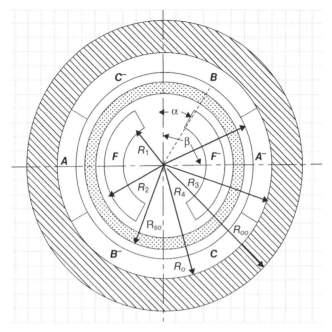

Figure 4.2   Schematic of cross section of a two-pole synchronous machine

motor and generator because the design equations are identical for both types of machines.

4.3.1   Magnetic Circuit and Harmonic Components

Although other configurations are possible, the synchronous machine configuration selected here has stationary armature winding, and field excitation winding and EM shield are on the rotor. Schematic cross section of a two-pole synchronous machine is shown in Figure 4.2. Description of various regions is provided in Table 4.1.

Analysis assumes that length of the machine is infinite along its z-axis. Thus the end effects of finite coils are neglected. It is possible make empirical allowance for these effects, which is discussed later. For the machine of Figure 4.2, it possible to describe field due to a two-pole field winding by Laplace's equation (4.1) in cylindrical coordinates [1], [2]. The z-axis is along the length (shaft) of the machine.

$$\frac{\delta^2 A_z}{\delta\rho^2} + \frac{1}{\rho} \cdot \frac{\delta A_z}{\delta\rho} + \frac{1}{\rho^2} \cdot \frac{\delta^2 A_z}{\delta\theta^2} = 0, \tag{4.1}$$

Table 4.1 Components in a two-pole machine cross section

Symbol	Description
F, F^-	Field winding, positive and negative current regions
A, A^-	Armature phase A winding, positive and negative current regions
B, B^-	Armature phase B winding, positive and negative current regions
C, C^-	Armature phase C winding, positive and negative current regions
R_1	Inside radius of field winding
R_2	Outside radius of field winding
R_3	Inside radius of armature winding
R_4	Outside radius of armature winding
R_{so}	Outside radius of EM shield
t_s	Radial thickness of EM shield
R_o	Inside radius of yoke
R_{oo}	Outside radius of yoke
α	half of field winding pole angle in electrical degree
β	Maximum field winding breath angle ($\leq$90 degree electrical)

where

A = vector potential,

ρ = radius,

θ = angle measured from midplane of a pole.

The solution of this equation for the vector potential A is of the form

$$A = \left(a \cdot \rho^{n \cdot p} + b \cdot \rho^{-n \cdot p}\right) \cdot \left[c \cdot \sin(n \cdot p \cdot \theta) + d \cdot \cos(n \cdot p \cdot \theta)\right]. \quad (4.2)$$

Radial and tangential field components are given by equations (4.3) and (4.4), respectively:

$$H_r = \frac{\delta A}{\delta \rho}, \quad (4.3)$$

$$H_\theta = \frac{1}{\rho} \cdot \frac{\delta A}{\delta \theta}. \quad (4.4)$$

The Fourier expansion of a current distribution in a thin cylindrical shell of the field winding of thickness ($d\rho$) and radius (ρ) has the form

$$K(\rho) = \frac{a_o}{2} + \sum_{n=1}^{\infty} \left[a_n \cos(n \cdot p \cdot \theta) + b_n \sin(n \cdot p \cdot \theta)\right]. \quad (4.5)$$

For a symmetrical distribution, the coefficients a_o and a_n in equation (4.5) are zero as shown below in equations (4.6) and (4.7):

$$a_o = \frac{1}{\pi} \cdot \int_{-\pi+\alpha}^{-\alpha} -J_f \cdot d\rho \cdot d(p \cdot \theta) + \frac{1}{\pi} \cdot \int_{\alpha}^{\pi-\alpha} J_f \cdot d\rho \cdot d(p \cdot \theta) = 0, \quad (4.6)$$

$$a_n = \frac{1}{\pi} \cdot \int_{-\pi+\alpha}^{-\alpha} -J_f \cos(n \cdot p \cdot \theta) \cdot d\rho \cdot d(p \cdot \theta)$$

$$+ \frac{1}{\pi} \cdot \int_{\alpha}^{\pi-\alpha} J_f \cos(n \cdot p \cdot \theta) \cdot d\rho \cdot d(p \cdot \theta) = 0. \quad (4.7)$$

The coefficient b_n reduces to

$$b_n = \frac{1}{\pi} \cdot \int_{-\pi+\alpha}^{-\alpha} -J_f \sin(n \cdot p \cdot \theta) \cdot d\rho \cdot d(p \cdot \theta)$$

$$+ \frac{1}{\pi} \cdot \int_{\alpha}^{\pi-\alpha} J_f \sin(n \cdot p \cdot \theta) \cdot d\rho \cdot d(p \cdot \theta),$$

$$b_n = \frac{4 J_f \cdot dr}{n \cdot \pi} \cos(n \cdot p \cdot \alpha), \quad (4.8)$$

and for the field winding, equation (4.5) reduces to

$$K(\rho) = \sum_{n\_odd} \left[\frac{4 J_f \cdot d\rho}{n \cdot \pi} \cos(n \cdot p \cdot \alpha) \cdot \sin(n \cdot p \cdot \theta) \right]. \quad (4.9)$$

The vector potential (A) is now given by

$$A = \sum_{n\_odd} \left[\frac{4 J_f \cdot d\rho}{n \cdot \pi} \cos(n \cdot p \cdot \alpha) \cdot \sin(n \cdot p \cdot \theta) \cdot \left(a \cdot \rho^{n \cdot p} + b \cdot \rho^{-n \cdot p} \right) \right]. \quad (4.10)$$

After equations (4.3), (4.4) and (4.10) are combined, radial and tangential field components are given by

$$H_r = \sum_{n\_odd} \left[\frac{4 J_f \cdot d\rho}{n \cdot \pi} \cos(n \cdot p \cdot \alpha) \cdot \sin(n \cdot p \cdot \theta) \right] \cdot n \cdot p \cdot \left(a \cdot \rho^{n \cdot p-1} - b \cdot \rho^{-n \cdot p-1} \right),$$

$$(4.11)$$

$$H_\theta = \sum_{n\_odd} \left[\frac{-4 J_f \cdot d\rho}{n \cdot \pi} \cos(n \cdot p \cdot \alpha) \cdot \cos(n \cdot p \cdot \theta) \right] \cdot n \cdot p \cdot \left(a \cdot \rho^{n \cdot p-1} - b \cdot \rho^{-n \cdot p-1} \right).$$

$$(4.12)$$

By applying appropriate boundary conditions for ferrous or conductive shields, various coefficients in equations (4.11) and (4.12) can be determined as reported elsewhere [1,2]. There are two-sets of equations: one for a machine with any number of pole-pairs (p) and another for a machine with $p = 2$.

Field due to Field Winding Current

Field due to Field Winding Current   Below are field components for a machine with any number of pole-pairs except 2.

Radial and tangential field components in the region at angular location ϕ and radius (ρ) ($\rho < R_1$) of the field winding are given by (4.13) and (4.14), respectively. Coefficient C_{fi} is given by (4.15).

$$B_r(\rho, n) = C_{fi} \cdot \left[\frac{R_2^{-n \cdot p+2} - R_1^{-n \cdot p+2}}{-n \cdot p + 2} + I \cdot \left[\frac{R_2^{n \cdot p+2} - R_1^{n \cdot p+2}}{(n \cdot p + 2) \cdot R_o^{2 \cdot n \cdot p}} \right] \right] \tag{4.13}$$
$$\cdot \rho^{n \cdot p-1} \cos(n \cdot p \cdot \phi),$$

$$B_\theta(\rho, n) = -C_{fi} \cdot \left[\frac{R_2^{-n \cdot p+2} - R_1^{-n \cdot p+2}}{-n \cdot p + 2} + I \cdot \left[\frac{R_2^{n \cdot p+2} - R_1^{n \cdot p+2}}{(n \cdot p + 2) \cdot R_o^{2 \cdot n \cdot p}} \right] \right] \tag{4.14}$$
$$\cdot \rho^{n \cdot p-1} \sin(n \cdot p \cdot \phi),$$

$$C_{fi} = \frac{\mu_0 \cdot J_f \cdot 2[\cos(\alpha \cdot n \cdot p) - \cos(\beta \cdot n \cdot p)]}{\pi \cdot n}, \tag{4.15}$$

where
B_r = radial field component,
B_θ = tangential field component,
J_f = overall winding current density,
n = odd harmonic number ($n = 1, 3, 5, \ldots$),
ϕ = angle measured from the pole axis,
I = +1 when shield at radius R_0 is ferrous,
I = −1 when shield at radius R_0 is conductive,
I = 0 when no shield is employed.

Radial and tangential field components within the winding region where ($R_1 < \rho < R_2$) are given by (4.16) and (4.17), respectively. Coefficient C_{fw} is given by (4.18).

$$B_r(\rho, n) = C_{fw} \cdot \left[\frac{-2 \cdot n \cdot p + (2 + n \cdot p)(R_2/\rho)^{-n \cdot p + 2} - (2 - n \cdot p) \cdot (R_1/\rho)^{n \cdot p + 2}}{2 - n \cdot p} \cdot \rho \right.$$

$$\left. + I \cdot \frac{R_2^{n \cdot p + 2} - R_1^{n \cdot p + 2}}{R_o^{2 \cdot n \cdot p}} \cdot \rho^{n \cdot p - 1} \right] \cdot \cos(n \cdot p \cdot \phi), \tag{4.16}$$

$$B_\theta(\rho, n) = C_{fw} \cdot \left[\frac{4 - (2 + n \cdot p)(R_2/\rho)^{-n \cdot p + 2} - (2 - n \cdot p) \cdot (R_1/\rho)^{n \cdot p + 2}}{2 - n \cdot p} \cdot \rho \right.$$

$$\left. - I \cdot \frac{R_2^{n \cdot p + 2} - R_1^{n \cdot p + 2}}{R_o^{2 \cdot n \cdot p}} \cdot \rho^{n \cdot p - 1} \right] \cdot \sin(n \cdot p \cdot \phi), \tag{4.17}$$

$$C_{fw} = \frac{\mu_0 \cdot J_f \cdot 2[\cos(\alpha \cdot n \cdot p) - \cos(\beta \cdot n \cdot p)]}{\pi \cdot n \cdot (n \cdot p + 2)}; \tag{4.18}$$

Radial and tangential field components outside the winding region where $(\rho > R_2)$ are given by (4.19) and (4.20), respectively. Coefficient C_{fo} is given by (4.21).

$$B_r(\rho, n) = C_{fo} \cdot \left(R_2^{2 + n \cdot p} - R_1^{2 + n \cdot p} \right) \cdot \left(\rho^{-n \cdot p - 1} + I \cdot \frac{\rho^{n \cdot p - 1}}{R_o^{2 \cdot n \cdot p}} \right) \cdot \cos(n \cdot p \cdot \phi), \tag{4.19}$$

$$B_\theta(\rho, n) = C_{fo} \cdot \left(R_2^{2 + n \cdot p} - R_1^{2 + n \cdot p} \right) \cdot \left(\rho^{-n \cdot p - 1} - I \cdot \frac{\rho^{n \cdot p - 1}}{R_o^{2 \cdot n \cdot p}} \right) \cdot \sin(n \cdot p \cdot \phi), \tag{4.20}$$

$$C_{fo} = \frac{\mu_0 \cdot J_f \cdot 2[\cos(\alpha \cdot n \cdot p) - \cos(\beta \cdot n \cdot p)]}{\pi \cdot n \cdot (n \cdot p + 2)}. \tag{4.21}$$

Below are field components for a machine with pole-pairs equal to 2 (i.e., number of poles = 4). Radial and tangential field components in the region where radius (ρ) is less than the inside radius (R_1) of the field winding are given by (4.22) and (4.23), respectively:

$$B_r(\rho, n) = C_{fi} \cdot \left[\ln\left(\frac{R_2}{R_1}\right) + I \cdot \left[\frac{R_2^{2 \cdot n + 2} - R_1^{2 \cdot n + 2}}{(2 \cdot n + 2) \cdot R_o^{4 \cdot n}} \right] \right] \cdot \rho^{n \cdot 2 - 1} \cos(n \cdot 2 \cdot \phi), \tag{4.22}$$

$$B_\theta(\rho, n) = -C_{fi} \cdot \left[\ln\left(\frac{R_2}{R_1}\right) + I \cdot \left[\frac{R_2^{n \cdot 2 + 2} - R_1^{n \cdot 2 + 2}}{(n \cdot 2 + 2) \cdot R_o^{4 \cdot n}} \right] \right] \cdot \rho^{n \cdot 2 - 1} \sin(n \cdot 2 \cdot \phi). \tag{4.23}$$

Radial and tangential field components within the winding region where $(R_1 < \rho < R_2)$ are given by (4.24) and (4.25), respectively:

$$B_r(\rho, n) = C_{fw} \cdot \left[\left[1 - \left(\frac{R_1}{\rho} \right)^{2 \cdot n + 2} + (2 \cdot n + 2) \cdot \ln \left(\frac{R_2}{\rho} \right) \right] \right.$$
$$\left. + \mathrm{I} \cdot \frac{R_2^{n \cdot 2 + 2} - R_1^{n \cdot 2 + 2}}{R_o^{4 \cdot n}} \cdot \rho^{n \cdot 2 - 2} \right] \cdot \rho \cos(n \cdot 2 \cdot \phi), \tag{4.24}$$

$$B_\theta(\rho, n) = C_{fw} \cdot \left[-\left[1 - \left(\frac{R_1}{\rho} \right)^{2 \cdot n + 2} - (2 \cdot n + 2) \cdot \ln \left(\frac{R_2}{\rho} \right) \right] \right.$$
$$\left. - \mathrm{I} \cdot \frac{R_2^{n \cdot 2 + 2} - R_1^{n \cdot 2 + 2}}{R_o^{4 \cdot n}} \cdot \rho^{n \cdot 2 - 2} \right] \cdot \rho \sin(n \cdot 2 \cdot \phi). \tag{4.25}$$

Radial and tangential field components outside the winding region where ($\rho > R_2$) are given by (4.26) and (4.27), respectively:

$$B_r(\rho, n) = C_{fo} \cdot \left(R_2^{2 + n \cdot 2} - R_1^{2 + n \cdot 2} \right) \cdot \left(\rho^{-n \cdot 2 - 1} + \mathrm{I} \cdot \frac{\rho^{n \cdot 2 - 1}}{R_o^{4 \cdot n}} \right) \cdot \cos(n \cdot 2 \cdot \phi), \tag{4.26}$$

$$B_\theta(\rho, n) = C_{fo} \cdot \left(R_2^{2 + n \cdot 2} - R_1^{2 + n \cdot 2} \right) \cdot \left(\rho^{-n \cdot 2 - 1} - \mathrm{I} \cdot \frac{\rho^{n \cdot 2 - 1}}{R_o^{4 \cdot n}} \right) \cdot \sin(n \cdot 2 \cdot \phi). \tag{4.27}$$

These equations (4.13 through 4.27) can be used for calculating field components B_r and B_θ due to field winding with a current density J_f in any region of Figure 4.2. Although the figure only shows a two-pole machines, the analysis presented in this section is applicable for any machine with p pole-pairs. The general nature of equations (4.13) through (4.27) is also identical for other winding—the only change is in the constants C (i.e., C_{fi}, C_{fw}, and C_{fo}). Field components for each harmonic can be calculated by select a value for n ($= 1, 3, 5, 7, \ldots$). Components due to different harmonics can be added to obtain total field at a desired point. Field components are needed for determining current-carrying capability of an HTS wire and for calculating AC losses.

Field due to Armature Winding Current   The field components due to armature current density J_a are also similar to equations (4.13) through (4.27) with the following changes:

- Replace R_1 and R_2 with R_3 and R_4, respectively.
- Replace coefficients C_{fi}, C_{fw}, and C_{fo} with C_{ai}, C_{aw}, and C_{ao}, respectively. Formulas for these coefficients are given by (4.28), (4.29), and (4.30):

$$C_{ai} = \frac{\mu_0 \cdot J_a \cdot 2 \cdot [\cos(\pi \cdot n \cdot p/3) - \cos(\beta \cdot n \cdot p)]}{\pi \cdot n}, \tag{4.28}$$

$$C_{aw} = \frac{\mu_0 \cdot J_a \cdot 2 \cdot [\cos(\pi \cdot n \cdot p/3) - \cos(\beta \cdot n \cdot p)]}{\pi \cdot n \cdot (n \cdot p + 2)}, \tag{4.29}$$

$$C_{ao} = \frac{\mu_0 \cdot J_a \cdot 2 \cdot [\cos(\pi \cdot n \cdot p/3) - \cos(\beta \cdot n \cdot p)]}{\pi \cdot n \cdot (n \cdot p + 2)}. \tag{4.30}$$

Field due to Shield Winding Current   The field components due to shield current density J_s, as given by equation (4.31), are also similar to equations (4.13) through (4.27) with the following changes:

- Replace R_1 and R_2 with R_{si} and R_{so}, respectively
- Replace coefficients C_{fi}, C_{fw}, and C_{fo} with C_{si}, C_{sw}, and C_{so}, respectively. Formulas for these coefficients are given by (4.32), (4.33), and (4.34).

Assuming that the shield winding has turns distributed sinusoidally but same current flows through each turn, the peak current density is then given by (4.21).

$$J_s = \frac{N_s \cdot I_s}{R_{so}^2 - R_{si}^2}. \tag{4.31}$$

The coefficients for field calculation are

$$C_{si} = \frac{\mu_0 \cdot J_s}{2}, \tag{4.32}$$

$$C_{sw} = \frac{\mu_0 \cdot J_s}{2 \cdot (n \cdot p + 2)}, \tag{4.33}$$

$$C_{so} = \frac{\mu_0 \cdot J_s}{2 \cdot (n \cdot p + 2)}. \tag{4.34}$$

Equations (4.13) through (4.34) can be used for calculating radial and tangential field components at any location in Figure 4.2 for each of the three windings, namely field, armature, and shield. All winding field components can be added to get the total radial and tangential fields at a desired location. Field components are needed for calculating current-carrying capability of HTS wire and the losses in it. Most currently available HTS wires are in tape form and have critical currents that are strongly dependent on the orientation of field with respect to the broad or narrow face of the wire. Thus it is necessary to calculate radial and tangential field components with respect to the HTS wire.

4.3.2 Parameter Calculations

Once magnetic field distribution due to all windings is known, each winding's self- and mutual inductances could be calculated with equations provided in this section. These equations utilize the finite radial thickness of each winding.

Self- and Mutual Inductances of Windings Self- and mutual inductances for various windings can be calculated by following the general approach discussed by Bumby [3]. However, the inductance formulas provided below were obtained by integrating flux linkages among windings of finite thickness (not just thin current sheets).

Armature Phase Winding Self-inductance (L_A) Self-inductance (L_A) of armature winding for nth space harmonic is given below with N_a turns per phase for $p \neq 2$ by (4.35) and for $p = 2$ by (4.36):

$$LA_a(n) = \frac{144 N_a^2 \cdot \mu_0 \cdot l \cos(n \cdot \pi/3)^2}{\pi^3 \cdot n^3 \cdot p \cdot (1-x^2)^2}$$

$$\cdot \left[\frac{(n \cdot p - 2) - (n \cdot p + 2) \cdot x^4 + 4 \cdot x^{n \cdot p + 2}}{(n \cdot p)^2 - 4} \right.$$

$$\left. + I \cdot 2 \cdot \left(\frac{R^4}{R_o} \right)^{2 \cdot n \cdot p} \cdot \frac{\left(1 - x^{n \cdot p + 2}\right)^2}{(n \cdot p + 2)^2} \right] \cdot k_w(n)^2, \tag{4.35}$$

$$L4_a(n) = \frac{144 N_a^2 \cdot \mu_0 \cdot l \cos(n \cdot \pi/3)^2}{\pi^3 \cdot n^3 \cdot p \cdot (1-x^2)^2}$$

$$\cdot \left[\frac{1-x^4}{4} + x^4 \ln(x) + I \cdot 2 \cdot \left(\frac{R^4}{R_o} \right)^{2 \cdot n \cdot p} \cdot \frac{\left(1 - x^{n \cdot p + 2}\right)^2}{(n \cdot p + 2)^2} \right] \cdot k_w(n)^2, \tag{4.36}$$

where

LA_a = armature self-inductance for a machine with $p \neq 2$,
$L4_a$ = armature self-inductance for a machine with $p = 2$,

$$x \quad = R_3/R_4,$$

N_a = number of turns/phase,

l = axial active length of armature winding,

μ_0 = permeability of free space = $4\pi \times 10^{-7}$ H/m,

$k_w(n)$ = winding factor for n^{th} space harmonic as given by (4.39).

The armature winding may employ coils with short pitch (<pole-pitch) and may have many coils in a phase belt. The pitch factor $k_p(n)$ and distribution factor $k_d(n)$ are defined below.

The pitch factor for a fractional pitch winding is given by (4.39) for the nth space harmonic.

$$k_p(n) = \cos\left(\frac{1-\xi}{2} \cdot \pi \cdot n\right), \tag{4.37}$$

where ξ is the coil pitch as a fraction of the pole-pitch. The distribution factor for a winding is given by

$$k_d(n) = \frac{\sin[N_c \cdot (\psi \cdot n)/2]}{N_c \sin(\psi \cdot n/2)}, \tag{4.38}$$

where

N_c = number of coils in a phase belt,

ψ = phase belt spread in electrical degrees (= $p \cdot$ mechanic degrees).

The winding factor $k_w(n)$ is given by

$$k_w(n) = k_p(n) \cdot k_d(n). \tag{4.39}$$

Field Winding Self-inductance (L_f) Self-inductance (L_f) of field winding for nth space harmonic is given below with N_f turns per pole for $p \neq 2$ by (4.40) and for $p = 2$ by (4.41):

$$LA_f(n) = \frac{16 \cdot N_f^2 \cdot \mu_0 \cdot l \cdot \left(\cos(\alpha \cdot n \cdot p)^2 - \cos(\beta \cdot n \cdot p)^2\right)}{\pi^3 \cdot n^3 \cdot p \cdot \left(1 - y^2\right)^2 \cdot [(\beta - \alpha) \cdot p/(\pi/2)]^2}$$
$$\cdot \left[\frac{(n \cdot p - 2) - (n \cdot p + 2) \cdot y^4 + 4 \cdot y^{n \cdot p + 2}}{(n \cdot p)^2 - 4}\right.$$
$$\left. + I \cdot 2 \cdot \left(\frac{R_2}{R_o}\right)^{2 \cdot n \cdot p} \cdot \frac{\left(1 - y^{n \cdot p + 2}\right)^2}{(n \cdot p + 2)^2}\right], \tag{4.40}$$

$$L4_f(n) = \frac{16 \cdot N_f^2 \cdot \mu_0 \cdot l \left(\cos(\alpha \cdot n \cdot p)^2 - \cos(\beta \cdot n \cdot p)^2 \right)}{\left[\pi^3 \cdot n^3 \cdot p \cdot (1 - y^2)^2 \right] \cdot [(\beta - \alpha) \cdot p/(\pi/2)]^2}$$

$$\cdot \left[\frac{1 - y^4}{4} + y^4 \ln(y) + I \cdot 2 \cdot \left(\frac{R_2}{R_o} \right)^{2 \cdot n \cdot p} \cdot \frac{(1 - y^{n \cdot p + 2})^2}{(n \cdot p + 2)^2} \right], \tag{4.41}$$

where

LA_f = field winding self-inductance for a machine with $p \neq 2$,
$L4_f$ = field winding self-inductance for a machine with $p = 2$,
y = R_1/R_2,
N_f = number of turns/pole,
l = axial active length of field winding.

Damper (Shield) Winding Self-inductance (L_s)   The superconducting field winding is likely to experience an AC field due to armature space harmonics, as well as any time harmonics imposed on the armature winding, when it is connected to an external source. These harmonic fields could induce excessive losses in the superconductor winding and its support structure. To minimize AC harmonic field experienced by the field winding, the superconductor winding is surrounded by a cylinder (shield) made of a conductive material like copper or aluminum. In a continuous shell, induced currents are free to flow in a nonuniform distribution throughout the shield. This shield is treated as a Fourier series of sinusoidally distributed damper windings (same as in [2]). The justification for this is that each space harmonic field will excite currents in this shield that are sinusoidally distributed. Self-inductance (L_s) of rotor electromagnetic shield for nth space harmonic is given below with N_s turns per pole for $p \neq 2$ by (4.42) and for $p = 2$ by (4.43):

$$LA_s(n) = \frac{N_s^2 \cdot \mu_0 \cdot l \cdot \pi}{4 \cdot n \cdot p} \cdot \frac{1}{\left[1 - (R_{s1}/R_{s2})^2 \right]^2}$$

$$\cdot \left[\frac{n \cdot p - 2 - (n \cdot p + 2) \cdot (R_{s1}/R_{s2})^4 + 4 \cdot (R_{s1}/R_{s2})^{n \cdot p + 2}}{(n \cdot p)^2 - 4} \right.$$

$$\left. + I \cdot 2 \cdot \left(\frac{R_{s2}}{R_o} \right)^{2 \cdot n \cdot p} \cdot \left[\frac{1 - (R_{s1}/R_{s2})^{n \cdot p + 2}}{(n \cdot p + 2)} \right]^2 \right], \tag{4.42}$$

$$L4_s(n) = \frac{N_s^2 \cdot \mu_0 \cdot l \cdot \pi}{4 \cdot n \cdot p} \cdot \frac{1}{\left[1 - (R_{s1}/R_{s2})^2\right]^2}$$

$$\cdot \left[\frac{1 - (R_{s1}/R_{s2})^4}{4} + \left(\frac{R_{s1}}{R_{s2}}\right)^4 \cdot \ln\left(\left(\frac{R_{s1}}{R_{s2}}\right)\right)\right.$$

$$\left. + I \cdot 2 \cdot \left(\frac{R_{s2}}{R_o}\right)^{2 \cdot n \cdot p} \cdot \left[\frac{1 - (R_{s1}/R_{s2})^{n \cdot p + 2}}{(n \cdot p + 2)}\right]^2\right], \tag{4.43}$$

where

LA_s = shield winding self-inductance for a machine with $p \neq 2$,
$L4_s$ = shield winding self-inductance for a machine with $p = 2$,
R_{s1} = inside radius of shield,
R_{s2} = outside radius of shield,
N_s = number of turns/pole (= 1 assumed).

Mutual Inductances Mutual inductances are needed among all windings. Furthermore, if a machine employs shield made of multiple shells of different materials, then each shell should be treated separately and mutual inductances among shells should also be calculated.

Armature Phase/Field Winding Mutual Inductance The mutual inductance between an armature phase and field winding is calculated by assuming that the field winding is excited, computing the flux linkages with one phase of the armature winding due to the magnetomotive force in the field winding, and dividing by the field winding current. The armature to field winding mutual inductance for nth space harmonic is given by (4.44) for $p \neq 2$ and by (4.45) for $p = 2$:

$$LA_{af}(n) = \left\{\frac{96 \cdot \mu_0 \cdot N_a \cdot N_f \cdot l \cdot [\cos(\alpha \cdot n \cdot p) - \cos(\beta \cdot n \cdot p)]}{\pi^3 \cdot n^3 \cdot p(1 - x^2) \cdot (1 - y^2) \cdot [(\beta - \alpha) \cdot p/(\pi/2)]} \cdot \cos[n \cdot (\pi/3)] \cdot \cos(n \cdot p \cdot \theta)}\left(\frac{R_2}{R_4}\right)^{n \cdot p}\right\}$$

$$\cdot (1 - y^{n \cdot p + 2}) \cdot \left[\frac{1 - x^{-n \cdot p + 2}}{4 - (n \cdot p)^2} + I \cdot \left(\frac{R_4}{R_o}\right)^{2 \cdot n \cdot p} \cdot \frac{1 - x^{n \cdot p + 2}}{(n \cdot p + 2)^2}\right] \cdot k_w(n)$$

$$\tag{4.44}$$

$$L4_{af}(n) = \left\{ \frac{96 \cdot \mu_0 \cdot N_a \cdot N_f \cdot l \cdot [\cos(n \cdot p \cdot \alpha) - \cos(n \cdot p \cdot \beta)]}{\pi^3 \cdot n^3 \cdot p \cdot (1 - x^2) \cdot (1 - y^2) \cdot [(\beta - \alpha) \cdot p/(\pi/2)]} \left(\frac{R_2}{R_4}\right)^{n \cdot p} \right\}$$

$$\cdot (1 - y^{n \cdot p + 2}) \cdot \left[\frac{-\ln(x)}{n \cdot p + 2} + I \cdot \left(\frac{R_4}{R_o}\right)^{2 \cdot n \cdot p} \cdot \frac{1 - x^{n \cdot p + 2}}{(n \cdot p + 2)^2} \right] \cdot k_w(n),$$

(4.45)

where

LA_{af} = mutual between armature and field windings for a machine with $p \neq 2$,

$L4_{af}$ = mutual between armature and field windings for a machine with $p = 2$.

Armature Phase/Shield Mutual Inductance The mutual inductance between armature phase and shield winding is calculated by assuming that the one phase of armature winding is excited, computing the flux linkages with an equivalent shield winding due to the magnetomotive force in the armature winding, and dividing by the armature winding current. The armature to shield winding mutual inductance for nth space harmonic is given by (4.46) for $p \neq 2$ and by (4.47) for $p = 2$:

$$LA_{as}(n) = \frac{12 \cdot \mu_0 \cdot N_a \cdot N_s \cdot l \cdot \cos(n \cdot \pi/3) \cdot \cos(n \cdot p \cdot \theta)}{\pi \cdot n^2 \cdot p \cdot (1 - x^2) \cdot \left[1 - (R_{s1}/R_{s2})^2\right]}$$

$$\left(\frac{R_{s2}}{R_4}\right)^{n \cdot p} \cdot \left[1 - \left(\frac{R_{s1}}{R_{s2}}\right)^{2 + n \cdot p}\right]$$

$$\cdot \left[\frac{1 - x^{2 - n \cdot p}}{4 - (n \cdot p)^2} + I \cdot \left(\frac{R_4}{R_o}\right)^{2 \cdot n \cdot p} \cdot \frac{1 - x^{n \cdot p + 2}}{(2 + n \cdot p)^2}\right] \cdot k_w(n), \qquad (4.46)$$

$$L4_{as}(n) = \frac{12 \cdot \mu_0 \cdot N_a \cdot N_s \cdot l \cdot \cos(n \cdot \pi/3) \cdot \cos(n \cdot p \cdot \theta)}{\pi \cdot n^2 \cdot p \cdot (1 - x^2) \cdot \left[1 - (R_{s1}/R_{s2})^2\right]}$$

$$\left(\frac{R_{s2}}{R_4}\right)^{n \cdot p} \cdot \left[1 - \left(\frac{R_{s1}}{R_{s2}}\right)^{2 + n \cdot p}\right]$$

$$\cdot \left[\frac{-\ln(x)}{2 + n \cdot p} + I \cdot \left(\frac{R_4}{R_o}\right)^{2 \cdot n \cdot p} \cdot \frac{1 - x^{n \cdot p + 2}}{(2 + n \cdot p)^2}\right] \cdot k_w(n), \qquad (4.47)$$

where

LA_{as} = mutual between armature and shield windings for a machine with $p \neq 2$,

$L4_{as}$ = mutual between armature and shield windings for a machine with $p = 2$.

Field Winding/Shield Mutual Inductance   The mutual inductance between field and shield windings is calculated by assuming that the field winding is excited, computing the flux linkages with an equivalent shield winding due to the magnetomotive force in the field winding, and dividing by the field winding current. The field to shield winding mutual inductance for nth space harmonic is given by (4.48) for $p \neq 2$ and by (4.49) for $p = 2$:

$$LA_{fs}(n) = \frac{\left[4 \cdot \mu_0 \cdot N_f \cdot N_s \cdot l \cdot [\cos(n \cdot p \cdot \alpha) - \cos(n \cdot p \cdot \beta)] \cdot (1 - y^{n \cdot p + 2})\right] \cdot (R_2/R_{s2})^{n \cdot p}}{\pi \cdot n^2 \cdot p \cdot [(\beta - \alpha) \cdot p/(\pi/2)] \cdot (1 - y^2) \cdot \left[1 - (R_{s1}/R_{s2})^2\right]}$$
$$\cdot \left[\frac{1 - (R_{s1}/R_{s2})^{2 - n \cdot p}}{4 - (n \cdot p)^2} + I \cdot \frac{(R_{s2}/R_o)^{2 \cdot n \cdot p} \cdot \left[1 - (R_{s1}/R_{s2})^{2 + n \cdot p}\right]}{(2 + n \cdot p)^2}\right],$$

$$(4.48)$$

$$L4_{fs}(n) = \frac{\left[4 \cdot \mu_0 \cdot N_f \cdot N_s \cdot l \cdot [\cos(n \cdot p \cdot \alpha) - \cos(n \cdot p \cdot \beta)] \cdot (1 - y^{n \cdot p + 2})\right] \cdot (R_2/R_{s2})^{n \cdot p}}{\pi \cdot n^2 \cdot p \cdot [(\beta - \alpha) \cdot p/(\pi/2)] \cdot (1 - y^2) \cdot \left[1 - (R_{s1}/R_{s2})^2\right]}$$
$$\cdot \left[\frac{-\ln(R_{s1}/R_{s2})}{2 + n \cdot p} + I \cdot \frac{(R_{s2}/R_o)^{2 \cdot n \cdot p} \cdot \left[1 - (R_{s1}/R_{s2})^{2 + n \cdot p}\right]}{(2 + n \cdot p)^2}\right],$$

$$(4.49)$$

where

LA_{fs} = mutual between field and shield windings for a machine with $p \neq 2$,

$L4_{fs}$ = mutual between field and shield windings for a machine with $p = 2$.

Shield 1 / Shield 2 Mutual Inductance   The mutual inductance between any two shield windings is calculated by assuming that shield

1 winding is excited, computing the flux linkages with an equivalent shield 2 winding due to the magnetomotive force in the shield 1 winding, and dividing by the shield 1 winding current. The shield 1 to shield 2 winding mutual inductance for nth space harmonic is given by (4.50) for $p \neq 2$ and by (4.51) for $p = 2$:

$$MS(n) = \frac{\mu_0 \cdot \pi \cdot l}{2 \cdot n \cdot p} \cdot \frac{(R_{t2}/R_{s2})^{n \cdot p} \cdot \left[1 - (R_{t1}/R_{t2})^{2+n \cdot p}\right]}{\left[1 - (R_{s1}/R_{s2})^2\right] \cdot \left[1 - (R_{t1}/R_{t2})^2\right]}$$
$$\cdot \left[\frac{1 - (R_{s1}/R_{s2})^{2-n \cdot p}}{4 - (n \cdot p)^2} + I \cdot \frac{(R_{s2}/R_o)^{2 \cdot n \cdot p} \cdot \left[1 - (R_{s1}/R_{s2})^{2+n \cdot p}\right]}{(2 + n \cdot p)^2}\right],$$

$$(4.50)$$

$$M4(n) = \frac{\mu_0 \cdot \pi \cdot l}{2 \cdot n \cdot p} \cdot \frac{(R_{t2}/R_{s2})^{n \cdot p} \cdot \left[1 - (R_{t1}/R_{t2})^{2+n \cdot p}\right]}{\left[1 - (R_{s1}/R_{s2})^2\right] \cdot \left[1 - (R_{t1}/R_{t2})^2\right]}$$
$$\cdot \left[\frac{-\ln(R_{s1}/R_{s2})}{2 + n \cdot p} + I \cdot \frac{(R_{s2}/R_o)^{2 \cdot n \cdot p} \cdot \left[1 - (R_{s1}/R_{s2})^{2+n \cdot p}\right]}{(2 + n \cdot p)^2}\right],$$

$$(4.51)$$

where

- MS = mutual inductance between shields for a machine with $p \neq 2$,
- $M4$ = mutual inductance between shields for a machine with $p = 2$,
- R_{s1}, R_{s2} = inside and outside radii of shield 1,
- R_{t1}, R_{t2} = inside and outside radii of shield 2.

Winding Resistances

Field Winding Resistance The field winding resistance is essentially zero. However, there is resistance of current leads and components in the exciter system. Typically a resistance of about $0.02\,\Omega$ could be used for the field winding.

Armature Winding Resistance Since the armature winding is assumed to have an overall current density of J_a, it is necessary to make an estimate of percent fraction of copper (space factor, λ) in the winding region. Depending on stator design approach the value of λ could vary between 0.15 and 0.3:

$$r_a = \frac{12 \cdot N_a^2 \cdot \rho_a \cdot l}{\pi \cdot \left[(R_4)^2 - (R_3)^2 \right] \cdot \lambda}, \tag{4.52}$$

where

r_a = resistance of armature/phase,

ρ_a = resistivity of copper at the operating temperature.

Resistance r_a must be increased by any eddy-current losses in the armature coils.

Shield Winding Resistance   Shield winding resistance is calculated using an assumption that turns are distributed sinusoidally over a pole-pitch. On this basis resistance of a shield winding is given as

$$r_s = \frac{N_s^2 \cdot \pi \cdot l \cdot \rho_{ss}}{4 \cdot R_{s2} \cdot (R_{s2} - R_{s1})}, \tag{4.53}$$

where

r_{as} = resistance of shield/phase,

ρ_{ss} = resistivity of shield material.

All resistances and inductances calculated above must be corrected for end connection of windings. An estimate of these will be discussed later.

Armature Shield Sizing   In the analysis above, armature is assumed to be shielded with magnetic iron ($I = 1$) or with a conductive shell ($I = -1$). This section describes how to calculate thickness of magnetic or conductive shields.

Magnetic Shield (or Iron Yoke)   This analysis assumes that the iron has infinite permeability. The inner radius of the iron shield is specified as R_o as shown in Figure 4.1. The outside radius of the iron shield is calculated on the basis of carrying all flux outside radius R_o when the field winding is excited with a field current necessary to generated rated voltage (on open-circuit) at machine terminals. Under load conditions the flux paths in the machine change significantly because of the armature reaction. However, to the first order, the total flux in the yoke will be about equal to the total flux on the open circuit (at rated voltage). During the open circuit the magnetic field to be carried by the yoke can be calculated by integrating the radial field, at radius R_o, produced by the field winding from the pole-axis to half a pole-pitch. This is given by

$$\Phi = \int_0^{\frac{\pi}{2p}} \sum_{n\_odd} (B_r(R_o, \phi, n) \cdot R_o \cdot l) d\phi, \tag{4.54}$$

where $B_r(R_o, \phi, n)$ is the radial field calculated with (4.19) at radius R_o and for odd space harmonics.

With a known working field (B_{yoke}) in the yoke, the radial thickness of the yoke (t_{yoke}) can be calculated from

$$t_{yoke} = \frac{\Phi}{l \cdot B_{yoke}}, \tag{4.55}$$

where l is the axial length of the machine. Thus the outside radius of the yoke is given by

$$R_{oo} = R_o + t_{yoke}. \tag{4.56}$$

The losses in the iron yoke can be estimated from yoke weight by using a loss value typical for laminations employed.

Conductive Shield   The inside radius of conductive shield is defined as R_o. The conductive shield is assumed to be infinitely conductive, but in reality it is made of a conductive material such as copper or aluminum. The purpose of the shield is to attenuate all the magnetic field outside of the radius R_o. Usually a shield thickness (t_s) equal to about 3 times the skin depth of shield material at fundamental frequency (f) of the machine is sufficient. The skin depth (Δ) and shield thickness can be calculated with (4.57) and (4.58), respectively:

$$\Delta = \sqrt{\frac{2 \cdot \rho_s}{\omega \cdot \mu_0}}, \tag{4.57}$$

where

ρ_s = resistivity of shield material,
ω = rotation frequency = $2\pi f$,
μ_0 = permeability of free space,

$$t_s = 3\Delta. \tag{4.58}$$

The maximum tangential field (B_{θ_max}) is obtained by combining (4.20) and (4.21) at $\rho = R_o$ is given in (4.59). The B_{θ_max} must be calculated using J_f that generated the rated voltage at machine terminals on open circuit.

$$B_{\theta\_max} = \frac{\mu_0 \cdot J_f \cdot 2 \cdot [\cos(\alpha \cdot n \cdot p) - \cos(\beta \cdot n \cdot p)]}{\pi \cdot n \cdot (n \cdot p + 2)}$$
$$\cdot \left(R_2^{2+n \cdot p} - R_1^{2+n \cdot p}\right) \cdot \left(2 \cdot R_o^{-n \cdot p - 1}\right). \tag{4.59}$$

The losses in the shield (per unit area) are given by

$$p_s = \rho_s \cdot \frac{B_{\theta\_max}^2}{2 \cdot \Delta}. \tag{4.60}$$

The total loss in the shield is given by

$$P_s = \left(B_{\theta\_max}\right)^2 \cdot \pi \cdot R_o \cdot l \cdot \sqrt{\frac{\omega \cdot \mu_0 \cdot \rho_o}{2}}. \tag{4.61}$$

4.3.3 Machine Terminal Parameters

Like any conventional synchronous machine, the superconducting machine must also be characterized with parameters as viewed from its terminals. This section calculates these parameters for a superconducting machine on the same basis as that for a conventional machine.

Machine Output Parameters   Machine terminal parameters can be calculated from the inductance and resistance values defined in previous sections. Armature phase current is

$$I_a = J_a \cdot \frac{\pi \cdot \left(R_4^2 - R_3^2\right)}{6 \cdot N_a}. \tag{4.62}$$

The field current is a function of the field winding's overall current density (J_f), which will change as a function of load. For a generator operating at a given lagging power factor, the J_f should be sufficient to support machine's terminal voltage and the voltage drop in resistance and synchronous reactance of the machine. With known J_f, the field current can be calculated from

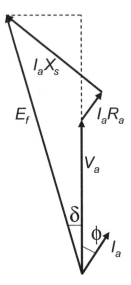

Figure 4.3   Vector diagram for a generator operating at lagging power factor

$$I_f = \frac{\pi}{2} \cdot \left(R_2^2 - R_1^2 \right) \cdot \left(\frac{\beta - \alpha}{\pi/(2 \cdot p)} \right) \cdot J_f. \tag{4.63}$$

Machine Steady-State Operation   A vector diagram for a generator relating terminal voltage with excitation voltage is provided in Figure 4.3. The parameters needed for completing the vector diagram can be obtained as explained below. Values of LA_a and LA_{af} are computed for the fundamental harmonic ($n = 1$).

R_a = resistance of armature winding calculated from (4.52),
X_s = synchronous reactance for a three-phase machine with $X_d = X_q$ is $1.5 \omega \cdot LA_a(n = 1)$,
f = frequency,
ω = $2\pi f$,
I_a = armature current calculated from (4.62),
I_f = field winding current calculated from (4.63),
L_{af} = mutual inductance between armature and field windings = $LA_{af}(n = 1)$,
E_f = field winding voltage = $\omega \cdot L_{af} \cdot I_f / \sqrt{2}$,
ϕ = power factor angle.

The load angle, δ, can be calculated from (4.64) using quantities defined above:

$$\sin(\delta) = \frac{I_a \cdot (X_s \cos(\phi) - R_a \sin(\phi))}{E_f}. \tag{4.64}$$

The machine terminal phase voltage is given by

$$V_a = E_f \cos(\delta) - I_a \cdot X_s \sin(\phi) - I_a \cdot R_a \cos(\phi). \tag{4.65}$$

The machine rating at its terminals is given by

$$VA = 3V_a I_a, \tag{4.66}$$

and its output power is given by

$$P = 3V_a I_a \cos(\phi). \tag{4.67}$$

Harmonic Components in Field Distribution   The voltage induced in the armature winding is directly proportional to the magnetic flux created by the field winding on the rotor and its interaction with the armature winding. If the field created by the field winding in the armature region has harmonics, then the voltage induced in the armature winding will have corresponding harmonic voltages. Usually harmonic

Harmonic Order	Fraction of Fundamental
1	1
3	0
5	$5.977 \cdot 10^{-4}$
7	$7.394 \cdot 10^{-6}$
9	0
11	$6.074 \cdot 10^{-8}$
13	$1.542 \cdot 10^{-10}$
15	0
17	$1.032 \cdot 10^{-11}$
19	$-3.381 \cdot 10^{-14}$

Figure 4.4   Harmonic content of field winding radial field at midpoint of armature winding

Harmonic Order	Fraction of Fundamental	
	Radial	Tangential
1	1	1
3	−0.057	0.116
5	$3.869 \cdot 10^{-4}$	$7.097 \cdot 10^{-3}$
7	$4.567 \cdot 10^{-5}$	$-1.153 \cdot 10^{-3}$
9	$-7.546 \cdot 10^{-5}$	$-4.378 \cdot 10^{-4}$
11	$4.55 \cdot 10^{-6}$	$-4.586 \cdot 10^{-5}$
13	$-8.935 \cdot 10^{-7}$	$1.027 \cdot 10^{-5}$
15	$5.559 \cdot 10^{-7}$	$4.827 \cdot 10^{-6}$
17	$-1.149 \cdot 10^{-8}$	$5.882 \cdot 10^{-7}$
19	$-2.729 \cdot 10^{-9}$	$-1.475 \cdot 10^{-7}$

Figure 4.5 Harmonic content of armature field at outer surface of EM shield on the rotor

components due to a superconducting winding in armature region are low, as shown in Figure 4.4 (an example design is discussed later in Section 4.3.5.) This figure shows that all harmonic components of radial field are negligible in the armature region, and as a result the voltage induced in the armature will be essentially a pure sine wave. This is far superior to that in a conventional machine in which harmonic content is usually large and corrective action is often needed (by using short-pitch coils) to minimize their effect on the generated voltage.

Likewise Figure 4.5 shows harmonic field radial and tangential components due to the phase-A armature current at the outer surface of the EM shield on the rotor. Again, these field distributions are far superior to those encountered in conventional machines. Only the third harmonic has a significant fraction. However, when fields of all three phases are combined, the effect of third harmonic cancels out. Low harmonic content of the armature field at the EM shield surface is important for the following reasons:

- Lower harmonic content induces less loss in the EM shield.
- Any fields due to time harmonic current flowing in armature coils (due to external sources) are also considerably attenuated.

For these reasons a superconducting machine with a warm EM shield is most suitable for synchronous motors operated from an electric grid or electric drive. Such a 5-MW, six-pole motor has been successfully tested [9] while fed from a variable speed drive (VSD).

Parameter Calculations   A superconducting machine must be characterized as a conventional machine in order to employ existing analysis tools for their integration and operation in an electric grid. Equations are provided in Section 4.3.2 for calculating resistance and inductance of individual circuits and mutual inductances among various circuits. Using (d-q)-axis representation, a conventional synchronous machine is usually characterized with the following parameters:

- Synchronous (x_d), transient (x_d') and subtransient (x_d'') reactances for d-axis.
- Armature short-circuit time constant (τ_a).
- Transient (τ_d') and subtransient (τ_d'') short-circuit time constants for d-axis.
- Transient (τ_{do}') and subtransient (τ_{do}'') open-circuit time constants for d-axis.
- Synchronous (x_q) and subtransient reactance (x_q'') for q-axis.
- Subtransient short-circuit time constants (τ_q'') for q-axis.
- Subtransient open-circuit time constants (τ_{qo}'') for q-axis.

These parameters could be calculated using individual circuit parameters as described below.

Synchronous reactance on d-axis is given by equation (4.68) for a N_{ph} machine.

$$x_d = 0.5 N_{ph} \cdot \omega \cdot L_a, \tag{4.68}$$

where

N_{ph} = number of phases in armature winding,
L_a = self-inductance of phase-A winding from equation (4.35) or (4.36).

Transient reactance on d-axis (x_d') is given by*

$$x_d' := \frac{N_{ph}}{2} \cdot \left(x_a - \frac{x_{af}^2}{x_f} \right). \tag{4.69}$$

*All reactances (X) and inductances (L) are related as $x = \omega \cdot L$, where ω is the nominal rotational frequency of the machine.

Subtransient reactance on d-axis (x_d'') is given by

$$x_d'' := \frac{N_{ph}}{2} \cdot \left(x_a - \frac{x_s \cdot x_{af}^2 - 2 \cdot x_{af} \cdot x_{as} \cdot x_{fs} + x_f \cdot x_{as}^2}{x_s \cdot x_f - x_{fs}^2} \right). \tag{4.70}$$

Since there is no saliency in a superconducting machine, synchronous reactance on the q-axis is the same as that on the d-axis:

$$x_q = x_d. \tag{4.71}$$

Due to absence of field winding on the q-axis, transient reactance is not defined for the q-axis. The subtransient reactance (x_q'') on the q-axis is given by

$$x_q'' := \frac{N_{ph}}{2} \cdot \left(x_a - \frac{x_{as}^2}{x_s} \right). \tag{4.72}$$

Armature short-circuit time constant (τ_a) is given by

$$\tau_a := \frac{1}{\omega \cdot r_a} \left[\frac{2}{(1/x_d'') + (1/x_q'')} \right]. \tag{4.73}$$

The d-axis transient open-circuit (τ_{do}') and short-circuit (τ_d') time constants are given by equation (4.74) and equation (4.75), respectively:

$$\tau_{do}' := \frac{x_f}{\omega \cdot r_f}, \tag{4.74}$$

$$\tau_d' := \frac{1}{\omega \cdot r_f} \cdot \left(x_f - \frac{x_{af}^2}{x_a} \right). \tag{4.75}$$

The d-axis subtransient open-circuit (τ_{do}'') and short-circuit (τ_d'') time constants are given by equation (4.76) and equation (4.77), respectively:

$$\tau_{do}'' := \frac{1}{\omega \cdot r_s} \cdot \left(x_s - \frac{x_{fs}^2}{x_f} \right), \tag{4.76}$$

$$\tau_d'' := \frac{1}{\omega \cdot r_s} \cdot \left(x_s - \frac{x_f \cdot x_{as}^2 - 2 \cdot x_{as} \cdot x_{af} \cdot x_{fs} + x_a \cdot x_{fs}^2}{x_a \cdot x_f - x_{af}^2} \right). \tag{4.77}$$

The q-axis subtransient open-circuit (τ_{qo}'') and short-circuit (τ_q'') time constants are given by equation (4.78) and equation (4.79), respectively:

$$\tau_{qo}'' := \frac{x_s}{\omega \cdot r_s},$$ (4.78)

$$\tau_q'' := \frac{1}{\omega \cdot r_s} \cdot \left(x_s - \frac{x_{as}^2}{x_a} \right).$$ (4.79)

Machine Parameters in Per-Unit   Electric machines are generally characterized in terms of equivalent circuits, which in turn are used for system design and analysis. A most commonly accepted equivalent circuit assumes that mutual reactance among armature, field, and EM shield circuits are identical. This assumption is nearly accurate for conventional iron-core machines but is not valid for superconducting machines employing air-core winding for armature, field, and EM shield circuits. Still, most commercial computer codes used for rotating machine integration with an electric grid or power electronics employ a representation of a machine using equivalent circuits with per-unit parameters. An equivalent circuit assumes a common mutual reactance among all circuits on each axis in a machine. This assumption produces results within 5% accuracy which is usually acceptable for most applications. However, it enables existing design tools to be employed for the design and integration of synchronous machines in electrical grid or other systems. Figures 4.6 and 4.7 show equivalent circuits for the d- and q-axes, respectively. Common mutual reactances (x_{ad} and x_{aq}) are arbitrarily selected as defined by

$$\left. \begin{aligned} x_{ad} &= x_d - 0.5x_d'' \quad \text{and} \quad x_{aq} = x_{ad.} \\ \text{and} \quad x_\lambda &= 0.5x_d'' \end{aligned} \right\}$$ (4.80)

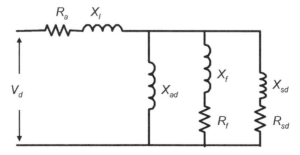

Figure 4.6   Equivalent circuit on d-axis for a synchronous machine

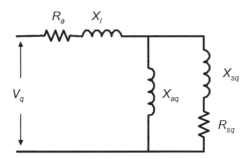

Figure 4.7   Equivalent circuit on *q*-axis for a synchronous machine

Per-unit (pu) base quantities for stator, field, and EM shield are defined in equations (4.81), (4.82) and (4.83), respectively:

PER-UNIT BASE QUANTITIES

Base voltage $\quad V_B = V_a \cdot \sqrt{2}$

Base current $\quad I_B = I_a \cdot \sqrt{2}$

Base power $\quad P_B = 1.5 \cdot V_B \cdot I_B$

Base impedance $\quad Z_B = \dfrac{V_B}{I_B}$ $\qquad\qquad\qquad$ (4.81)

Base field current $\quad I_{fB} = \dfrac{x_{ad}}{x_{af}} \cdot I_B$

Base field voltage $\quad V_{fB} = \dfrac{P_B}{I_{fB}}$

Base impedance $\quad Z_{fB} = \dfrac{V_{fB}}{I_{fB}}$ $\qquad\qquad\qquad$ (4.82)

Base EMS current $\quad I_{sB} = \dfrac{x_{ad}}{x_{as}} \cdot I_B$

Base EMS voltage $\quad V_{sB} = \dfrac{P_B}{I_{sB}}$

Base EMS impedance $\quad Z_{sB} = \dfrac{V_{sB}}{I_{sB}}$ $\qquad\qquad\qquad$ (4.83)

Except for x_{ad}, all reactance values in Figure 4.6 are leakage reactances. The per-unit values for various quantities follow. All per-unit (pu) quantities are in uppercase letter.

$$X_{ad} = X_{aq} = \frac{x_{ad}}{Z_B}. \tag{4.84}$$

Per-unit armature winding resistance (R_a) and leakage reactance (X_l) are

$$R_a = \frac{r_a}{Z_B}, \tag{4.85}$$

$$X_l = \frac{x_l}{Z_B}. \tag{4.86}$$

Per-unit field winding resistance (R_f) and leakage reactance (X_f) are

$$R_f = \frac{r_f}{Z_{fB}}, \tag{4.87}$$

$$X_f = \frac{x_f}{Z_{fB}} - X_{ad}. \tag{4.88}$$

Per-unit field EM shield resistances (R_s) and leakage reactances (X_s) are

$$R_{sd} = R_{sq} = \frac{r_s}{Z_{sB}}, \tag{4.89}$$

$$X_{sd} = X_{sq} = \frac{x_s}{Z_{sB}} - X_{ad}. \tag{4.90}$$

Since there is no saliency in superconducting machines, the armature and shield per-unit quantities on the q-axis are equal to those on the d-axis.

Machine's transient parameters (per-units) can now be defined as below using equations from Adkins [4].

For the d-axis synchronous reactance,

$$X_d := X_{ad} + X_l. \tag{4.91}$$

For the d-axis transient reactance,

$$X'_d := X_l + \frac{1}{(1/X_{ad}) + (1/X_f)}. \tag{4.92}$$

For the *d*-axis subtransient reactance,

$$X_d'' := X_l + \frac{1}{(1/X_{ad}) + (1/X_f) + (1/X_{sd})}. \tag{4.93}$$

For the *q*-axis transient reactance,

$$X_q := X_{aq} + X_l. \tag{4.94}$$

For the *q*-axis sub-transient reactance,

$$X_q'' := X_l + \frac{1}{(1/X_{aq}) + (1/X_{sq})}. \tag{4.95}$$

For the *d*-axis transient open-circuit time constant,

$$\tau_{do}' := \frac{X_{ad} + X_f}{\omega \cdot R_f}. \tag{4.96}$$

For the *d*-axis transient short-circuit time constant,

$$\tau_d' := \frac{1}{\omega \cdot R_f} \left[X_f + \frac{1}{(1/X_l) + (1/X_{ad})} \right]. \tag{4.97}$$

For the *d*-axis subtransient open-circuit time constant,

$$\tau_{do}'' := \frac{1}{\omega \cdot R_{sd}} \left[X_{sd} + \frac{1}{(1/X_{ad}) + (1/X_f)} \right] \tag{4.98}$$

For the *d*-axis subtransient short-circuit time constant,

$$\tau_d'' := \frac{1}{\omega \cdot R_{sd}} \cdot \left[X_{sd} + \frac{1}{(1/X_l) + (1/X_{ad}) + (1/X_f)} \right]. \tag{4.99}$$

For the *q*-axis subtransient open-circuit time constant,

$$\tau_{qo}'' := \frac{X_{aq} + X_{sq}}{\omega \cdot R_{sq}}. \tag{4.100}$$

For the q-axis subtransient short-circuit time constant,

$$\tau_q'' := \frac{1}{\omega \cdot R_{sq}} \cdot \left[X_{sq} + \frac{1}{(1/X_l) + (1/X_{aq})} \right]. \qquad (4.101)$$

4.4 DESIGN

This section discusses machine design issues relating to stator, rotor, and EM shield. An example 10-MW generator design is presented later in this section for demonstrating the use of equations developed earlier in this chapter. These equations are generic and can be utilized for designing a machine of any rating or any number of poles.

4.4.1 Stator Winding Design Issues

Stator Winding Configuration   Armature (stator) winding is usually quite similar to that for a conventional rotating synchronous or induction machine. Both single-layer and double-layer winding approaches are possible. Usually superconducting machines are designed with air-gap armature winding, namely without an iron tooth stator. However, it is possible to employ stator windings with magnetic iron teeth in superconducting machines. Using an iron-tooth stator usually limits the design of the superconducting field winding to low fields to avoid saturation of iron teeth. However, if the iron teeth are replaced with a nonmagnetic, nonconductive structure, then it is possible to operate the superconducting field winding at much higher magnetic field levels. The field level in stator winding region could be considerably higher than in an iron tooth stator. It is possible to design a superconducting machine with a field in the stator winding region ranging from 1 to 2T. If iron teeth are employed in such stators, with a slot-width/slot-pitch ratio of 0.5, then the theoretical field in the teeth would be in the range of 2 to 4T. A tooth field of this magnitude will cause high iron losses, impacting machine efficiency and presenting serious difficulties in removing heat from local hot spots in the teeth. The iron-teeth saturation will also result in high field leakage in the slot, and thus in high copper loss unless finely stranded conductors were used. Clearly, some other means and materials (nonconductive, nonmagnetic) must be employed to support the stator winding against the steady-state and short-circuit forces.

Removal of the iron teeth will permit additional copper area in the stator region, increasing the stator winding current loading. Since a machine's rating is a direct function of armature loading and the field in armature region due to superconducting field winding, a superconducting machine can be much more compact than a conventional machine of similar rating.

In a conventional iron-tooth stator, stator winding coils are subjected to only cross-slot fields (1D). These coils are constructed by employing thin copper conductors (thinner part faces the cross-slot field). However, when iron teeth are eliminated, the stator coils are subjected to full AC field varying in both radial and tangential directions (2D). Thus a coil turn in an air-gap winding must be made of small diameter strands for two reasons: (1) it is exposed to two components of alternating field (or a rotating flux wave), and (2) the field levels are higher (in a superconducting machine). Small-diameter strands must be individually insulated and transposed to limit circulating currents. Such conductors are commercially available and are known as "Litz" wire. However, the stator coils comprising a Litz wire yield a poor space factor compared to conventional machine coils. To some extent this poor space factor negates the advantage of space saving due to removal of iron teeth in an air-gap winding. Furthermore armature loading is strongly influenced by type of winding employed (with or without iron teeth) and type of cooling (air or liquid cooled). Synchronous machines have been successfully built and operated using both approaches. The design equations described in this book are for an air-core stator winding, without iron teeth. The machine designs involving iron teeth require nonlinear finite-element (F-E) analysis tools because of the saturation of iron teeth. Such stators are difficult to model analytically and are outside the scope of this book.

Stator Yoke (Shield) Configuration   The elimination of iron teeth from the stator winding region does not remove the necessity of containing the magnetic field of the machine, although it does permit some conceptual freedom of choice in the manner in which this is achieved. Three possible choices are listed below:

1. No shielding.
2. Laminated iron yoke.
3. Solid conducting yoke.

In Section 4.2.1 the magnetic field equations for all machine windings contain a constant I, which essentially defines the type of yoke (shielding) selection. The three possible choices for I corresponding to the three shielding options are as follows:

- $I = 0$ for no shielding.
- $I = +1$ for laminated iron yoke.
- $I = -1$ for solid conducting yoke.

The first option ($I = 0$) of no shielding is attractive in principle but difficult to envision, since a stray field can induce eddy-current heating in conductive metallic components around a machine. Also randomly distributed metallic components in vicinity of the machine can produce asymmetric forces within the machine stator and rotor. The biggest appeal of this option is that it can produce the most compact machine possible. In some outer space applications or for off-shore wind turbine generator applications this option could be of interest.

The second option ($I = +1$) is used in a majority of the superconducting machines constructed so far it. The laminated iron yoke that is characteristic of this option is manufactured in a manner similar to that currently employed in conventional machines, except the teeth are removed from punching. The principal disadvantage of this type of shield is that it is bulky though it does provide rigidity to the stator structure and damping of vibrations, both are major considerations in overall stator designs. This shield choice attenuates both AC and DC fields and enhances field (as is evident in field equations in Section 4.2.1) in the working region of the machine. Iron yoke thickness can be calculated using equation (4.55) for a given average field in the yoke.

The third option ($I = -1$) consist of enclosing the outside of the armature winding with a solid conductive shell made of copper, aluminum, or another high-conductivity metal. In such a shield, the shielding currents are skin-depth limited. This shield choice attenuates only the AC fields (i.e., when the rotor is rotating); no shielding is provided of the rotor field when the rotor is stationary. The shield must be remote from the stator winding to avoid excessive shield losses and to minimize the demagnetizing effect of the shield. The principle advantage of a solid conductive shield is that it can be of low weight compared to a magnetic yoke. Still some form of nonmetallic structure of high rigidity is required to support the stator winding, bear the weight of the machine, support short-circuit forces, and transmit the machine torque from the stator coils to the foundation. The thickness of the conductive shield and the eddy-current losses in it can be calculated using equations (4.57) and (4.58), respectively.

Stator Insulation The elimination of iron teeth between adjacent stator coils in a superconducting generator frees up space in the stator. This space can be utilized to achieve the following objectives:

- Thicker insulation to increase voltage capability of the machine.
- Additional copper to increase stator current loading.

All air-gap winding also presents a challenge. In air-cooled machines, iron teeth on either side of a coil help remove heat from a coil. In the absence of iron teeth, this capability is lost. Nonmagnetic thermal teeth have been postulated [5] to provide the cooling capability lost by the absence of iron teeth. Most air-gap windings considered for superconducting machines have adopted one of the following approaches:

- The coil conductor is cooled with a coolant (e.g., water) flowing in tubes buried in a coil cross section. The coolant is at the coil voltage, and special techniques are used for interfacing this fluid with cooling equipment at ground potential. These techniques are well established in industry because they are used in large conventional generator coils.
- It is possible to embed cooling tubes between adjacent coils. Suitable dielectric oil, such as FR3, could flow in them to cool coils. Several large machines have been built using this approach.

For large turbine generators, water-cooled bars are used for the stator winding [6].

4.4.2 Field Winding Design Issues

Under steady-state conditions, the superconducting rotor winding directly experiences the full-load torque and the centrifugal and internal magnetic forces. The mechanical design of the rotor must provide a structure capable of transmitting large torques but at the same time limit thermal conduction to the low-temperature region. In addition the differential contraction between the cryogenic- and ambient-temperature components of the rotor must be accommodated. The superconducting winding must be adequately cooled and thermally and mechanically stabilized. The superconducting winding must be protected from any AC harmonic field from the stator. These fields are caused by stator winding space harmonics, unbalanced stator currents, and time harmonics fed to the stator from external sources such as the grid or a power electronic drive employed to power a motor. Normally an electromagnetic (EM) shield in form of a solid cylinder made of a low-conductivity material is applied to enclosed the whole superconducting winding region. The design issues relating to this EM shield are discussed in the next section.

Under certain transient conditions, such as those caused by a fault on the grid, the field winding will experience magnetic field, temperature, and current changes simultaneously or independently. The magnetic field, temperature, and current excursions in the superconductor must be considered for each operating situation, to ensure that an adequate margin of safety is maintained below the critical values and to prevent a normal transition in the field winding.

Superconducting Field Winding The superconducting field winding is energized with DC and is designed to generate a magnetic field in the stator armature region. The stator windings experience this field as AC when the rotor rotates. In a generator this AC field produces voltage in the stator coils. The field experienced by the stator imposes following requirements on the field winding design:

- Only the fundamental component of the field generated by the field winding is useful for producing power.
- Any space harmonics created by the field winding will generate harmonic voltages in the stator coils and additional eddy-current heating in coils and other metallic components. Thus these space harmonics must be minimized.
- Field winding must be designed to achieve the highest possible current density in order to minimize space used on the rotor and to minimize use of expensive superconductor.
- Cooling of superconductor winding must be adequate to minimize the temperature rise that can occur under the transient and fault conditions that must be sustained by the machine.
- Mechanical stresses within the winding and associated support structure must be acceptable during the cool-down from the room temperature to the cryogenic operating temperature and during the warm-up to the room-temperature. The winding must also be capable of withstanding stresses under steady-state and fault conditions.

Many different approaches have been utilized by manufacturers for constructing superconducting field windings. A few major approaches are discussed below:

- Wind superconductor on salient poles directly to form field winding poles. In some cases the pole material has been a solid magnetic iron.
- Wind many racetrack-shaped coils and then assemble them to form poles. The pole body again can be magnetic or nonmagnetic.
- Wind superconductor directly into slots cut out in a solid magnetic iron, similar to the rotor of a conventional machine.

Almost in all cases, windings are epoxy impregnated for creating a monolithic structure to bear magnetic loads and for ease of cooling.

Over the 25-year span of 1970–1995, many superconducting machines were built [7] using low-temperature superconductors (niobium-titanium) operating at 4.5 K with liquid helium as a coolant. Most of these machines were successful but were not considered practical because of the cost of the refrigeration system and reliability issues. However, commercial availability of the high temperature superconductor (HTS) around 1995 changed the paradigm. These new HTS materials have the following attractive characteristics:

- HTS materials are superconducting at much higher temperatures, so practical devices can be constructed at temperatures ranging from 30 K to 77 K.
- The cost of refrigeration systems is lower at higher temperatures.
- HTS materials operating at higher temperatures have higher capability to absorb transient heating without transitioning to normal state.
- Refrigeration systems in modular form are reliable and are easy to operate.

Torque Tubes Superconducting windings experience the full rated torque of a machine. This torque in superconducting windings must be transferred to a warm shaft that interfaces with the power source (prime mover) for generators and with loads for motors. The component that transfers torque from cryogenic temperature field windings to the warm shaft is called a "torque tube." The torque tube transfers torque while minimizing heat conduction from the room-temperature shaft to the cryogenic environment on the rotor. For low-temperature machines, helium gas cooled torque tubes were employed to minimize the refrigeration load. These torque tubes were very complex. However, torque tubes are much simpler for HTS field windings because the penalty for removing losses from cryogenic temperature is lower.

Torque tubes used successfully in recent HTS machines were made of both metallic and nonmetallic composite materials.

Current Leads   The superconducting field winding must be supplied with DC from room-temperature power sources. The current leads that interface between the superconducting windings and the room-temperature power source also conduct thermally. Any thermal load conducted down leads must be removed by the refrigeration system. Usually current leads are made from brass or bronze. The cross section of a lead is optimally varied between the room-temperature and the cryogenic temperature end points. An optimally designed lead has thermal conduction equal to the I^2R loss in it. The process for designing such a lead has been discussed in Chapter 3 and is described elsewhere [8,9].

Cryostat   The superconducting field winding in its cryogenic environment must be protected from the heat radiated and conducted from room-temperature environment. The rules discussed in Chapter 3 for designing a cryostat also apply for the superconducting field's winding cryostat. Extra care is warranted because the cryostat rotates in a machine. Below are key features of a rotating machine cryostat:

- The room-temperature wall of the cryostat is mechanically tied to the shaft of the machine.
- The cold mass consisting of the field winding and its support structure is tied to the warm wall of the cryostat through torque tubes.
- In some cases an intermediate-temperature (between the cryogenic and room temperature) EM shield is employed to intercept any thermal loads from the stator caused by the harmonic fields.
- The space between the warm and cold walls of a cryostat is usually evacuated and filled with multi-layer insulation (MLI), as is used for minimizing the thermal radiation from the warm surfaces to cold surfaces.
- Besides torque tubes, any other mechanical connections between the cold mass and warm cryogenic wall must be optimally designed to minimize any thermal conduction into the cold environment of the field winding.

For HTS field windings, the intermediate thermal shield is usually not employed because it has only a minor impact on the total thermal load.

A superconducting field winding has the following four components of thermal load.

· Hysteresis and conduction losses in superconducting winding while carrying its rated current.
· Current lead loss.
· Thermal radiation and conduction from warm surfaces.
· Thermal conduction through torque tubes.

A good field winding design has these four thermal loads in about equal proportions.

4.4.3 Electromagnetic (EM) Shield Design Issues

A superconducting machine needs an EM shield to protect the superconducting field winding from the asynchronous fields produced by armature currents. Besides attenuating the AC fields, an EM shield must be capable of withstanding very large torques and forces experienced by it during a transient short-circuit fault. Large machines [10,11] built in recent years have utilized a room-temperature EM shield employed just outside of the warm wall of the rotor cryostat. The EM shield in this location experiences AC fields from stator winding and attenuates them sufficiently to minimize the thermal load in the cold region. This shield is also subjected to large torque and forces during transient faults such as short-circuit on or near machine terminals. An EM shield could be constructed from high-conductivity (HC) aluminum or copper to provide needed attenuation of AC fields. However, these HC metals are not sufficiently strong to bear fault torques and forces. It is also not advisable to increase the thickness of the aluminum or copper EM shields for this mechanical reason because a thicker shield lowers the damping torque during the hunting oscillations that follow a disturbance on the electric grid. Composite shields, initially proposed elsewhere [6], are better alternatives. A composite shield has an outer layer made of HC metal and backed with a stainless steel cylinder to provide the mechanical support needed during faults. The HC shield thickness is usually optimized to provide an adequate attenuation of the AC field and damping torque during the hunting oscillations, and the stainless steel cylinder is designed to withstand the fault forces. It is necessary to mechanically attach the HC shield to the stainless steel cylinder because the fault forces generated in the HC shield are borne by the stainless steel shell. The composite EM shield must be firmly attached to the shaft for transferring short-circuit torques.

4.4.4 Loss and Efficiency Calculations

Superconducting machines are usually more efficient than conventional machines. The major loss components of a superconducting machine are listed below:

- Usually stator windings are constructed using copper coils and therefore represent the largest loss component in the efficiency calculations. This component can be calculated using the armature resistance equation (4.52).
- The iron yoke loss is usually the next big loss component in a 60 Hz machine. This loss component can be estimated from loss/kg as a function of field and frequency supplied by lamination manufacturers.
- Friction and windage losses could also be significant. It is estimated using bearing types and cooling medium in the air gap.
- Superconducting field winding system loss (including input power to the refrigerator system) is usually the lowest loss component (about 2% of total loss). Losses in the superconducting winding itself are negligible. Most of the losses associated with the field winding come from input power into refrigeration system.

Loss distribution for a typical two-pole, 60-Hz generator is shown in Figure 4.8. At partial loads, all loss components except the stator loss

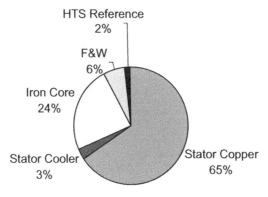

Figure 4.8   Loss distribution in a two-pole synchronous generator

essentially remain unchanged. The stator losses are low under part-load conditions because they are directly proportional to square of the load current. Thus, unlike conventional machines, a superconducting machine retains its high efficiency during partial loads. The generator represented in Figure 4.8 has a high efficiency at full-load (98.6%), and this efficiency is retained at partial loads up to less than a third of the rated load.

4.4.5 Example Design

Specifications Specifications are a starting point for the design of a motor or generator that must satisfy certain requirements. The overall specifications for a 10-MW generator are listed in Table 4.2. The generator was designed with the high temperature superconductor (HTS) of BSCCO-2223 (called 1G conductors). These conductors are currently being replaced with YBCO-123 coated conductors (called 2G condutors), but their engineering properties are not broadly available yet. Once YBCO properties become available, it would be a simple matter to replace BSCCO with YBCO in the design process.

Rating and Dimensions The output of a synchronous machine is based on the specific magnetic and electrical loadings, speed, air-gap diameter, and axial length. Low-speed machines have a large number of poles. Since the effective air gap in superconducting machines is very large, the leakage flux between adjacent poles increases as the number of poles is increased. This reduces the available flux that can link with the armature coils leading to a reduction in the output of the machine. Usually superconducting machines with a large number of poles are less attractive.

Table 4.2 Specifications for a 10 MW generator

Parameter	Unit	Value
Rating	MW	10
Line voltage	kV	13.8
Number of phases	—	3
Stator connection	—	Star
Frequency	Hz	60
Power factor at rated load	—	0.85 lagging
Speed	RPM	1200
Field winding (HTS)	—	Bi-2223
Number of poles	—	6

Specific Loading (Armature and Field)   The specific loading of an armature (A/m along bore perimeter) is a strong function of coil cooling. This is especially true for superconducting machines employing air-gap winding with thermally nonconductive teeth. An air-cooled winding will have a much lower armature-specific loading than a winding actively cooled with water or oil.

For a given superconductor, the field winding specific loading is a strong function of local magnetic field experienced by the winding and its operating temperature. The operating current of the superconducting wire is selected on the basis of losses in the superconductor and the ability to remove them safely during normal operation. The operating current is also influenced by the ability of the EM shield to protect the superconducting winding during a quench. Moreover, during a sudden short-circuit fault at or near machine terminals, the armature reaction current raises field current significantly. The increased field current could be equal to $\{(x_d - x'_d)/x'_d\}$ times pre-fault field current, and it decays with a transient short-circuit time constant (τ'_d) that is usually very long in superconducting machines. This enhanced field current could continue to flow for the duration of fault. The superconducting winding must therefore be designed to stay operating safely through such faults. As a rule of thumb, the superconductor operating current should be about three-quarters of the critical current of the winding at the rated operating temperature and magnetic field.

Main Dimensions   Because a superconducting machine air-gap winding has a large air-gap length, it is important to select a field winding pole pitch as large as possible. Typically a pole-pitch between 300 and 400 mm works fine. While it is possible to combine a conventional stator with magnetic iron teeth with a superconducting field winding, this combination requires that the air-gap field be kept below a value that does not saturate iron teeth. Such a machine is usually larger and heavier than a machine employing an air-gap armature winding in which the air-gap magnetic field is not limited by the saturation of iron teeth. A conventional laminated iron yoke is employed outside of armature winding to keep the field outside the machine at the same level as for nonsuperconducting conventional machines. It is also possible to employ a conductive shield (made of copper or aluminum) in place of magnetic iron yoke, but this option is normally not selected for the following two reasons:

  · It increases machine's overall diameter, but the machine's weight is less than that of the iron yoke shielded machine.

· The field outside the machine is quite high when the excited rotor is turning at less than its rated speed.

The armature winding's slot depth and slot pitch rules are similar to those of a conventional machine. However, in an air-gap winding the magnetic field varies in both radial and tangential directions, and this forces use of the Litz conductor for the armature coil. A Litz conductor is composed of many small diameter strands (insulated and fully transposed) for keeping the eddy-current losses low.

A practical design of a superconducting motor or generator has many complex engineering issues relating to mechanical, thermal, and electromagnetic designs. The process of designing the superconducting winding and its cooling system is also complex. It requires a great deal of design and analysis iterations before arriving at an acceptable design within constraints of material properties and physical limitation of an application. The example design described below illustrates the various issues that need consideration.

End Corrections   The parameter calculations in Section 4.2.2 are based on a 2D analysis that does not account for the effect of the coil end turns. The end-turn contribution is included by adjusting each parameter with certain empirical correction factors. These factors can be determined from 3D finite-element magnetic analysis or from measurements of a physical model, or both. The correction factors provided below enable reasonable estimates to be made for the example machine discussed in this section.

Armature Winding Corrections   The armature resistance and inductance are corrected using the following factors:

Multiply the resistance calculated using equation (4.52) with the following armature resistance correction factor:

$$AR_{CF} = 1 + 2 \cdot \frac{L_{EndTurn}}{l}, \tag{4.102}$$

where

l = active axial length of the machine,
$L_{EndTurn}$ = length of an end turn beyond end of a stator slot.

Multiply the inductances calculated using equations (4.35) and (4.36) with the following armature inductance correction factor:

$$AL_{CF} = 1 + \frac{(L_{EndTurn} - \text{EndLoop})}{3 \cdot l}, \qquad (4.103)$$

where

EndLoop = axial extension of knuckle of an armature coil.

Field Winding Corrections   For the field winding there is no correction factor for the resistance because the resistance of an HTS winding is essentially zero.

Multiply the field winding inductance calculated using equations (4.40) and (4.41) with the following field winding correction factor:

$$FL_{CF} = \left(\frac{R_2}{R_1} \right)^2. \qquad (4.104)$$

EM Shield Corrections   Multiply the shield resistance calculated using equation (4.53) with the following shield resistance correction factor:

$$SR_{CF} = 1 + \frac{2 \cdot \pi \cdot R_{so} \sin[(\alpha + \beta)/2]}{l}. \qquad (4.105)$$

The correction factor for the shield inductance is the same as that for the armature winding estimated from equation (4.103). Multiply the shield inductance calculated using equations (4.42) and (4.43) with the following shield winding correction factor:

$$SL_{CF} = AL_{CF}. \qquad (4.106)$$

Mutual Inductance between Armature Winding and Field Winding Corrections   The correction factor for the armature to field mutual inductance is tied by equation (4.107) below to the armature's self-inductance. Multiply the mutual inductance calculated using equations (4.44) and (4.45) with the following correction factor:

$$AF_{CF} = \sqrt{AL_{CF}}. \qquad (4.107)$$

Mutual Inductance between Armature Winding and EM Shield Corrections   The correction factor for the armature to shield mutual inductance is the same as that for the self-inductance of shield estimated using equation (4.106). Multiply the mutual inductance calculated using equations (4.46) and (4.47) with the following correction factor:

$$AS_{CF} = SL_{CF}. \tag{4.108}$$

Mutual Inductance between Field Winding and EM Shield Corrections   The correction factor for the armature to shield mutual inductance is the same as that for the self-inductance of the field winding found in equation (4.104). Multiply the mutual inductance calculated using equations (4.48) and (4.49) with the following correction factor:

$$FS_{CF} = FL_{CF}. \tag{4.109}$$

Mutual Inductance between EM Shield Shell Corrections   The correction factor for the mutual inductance between any two shield shells is the same as that for the self-inductance of the shield estimated using equation (4.105). Multiply the mutual inductance calculated using equations (4.50) and (4.51) with the following correction factor:

$$SS_{CF} = SL_{CF}. \tag{4.110}$$

The purpose of the correction factor definitions in this section is to illustrate how to use them for correcting various winding parameters. However, it is necessary to determine their validity for a given frame size of a machine. Once the correction factors are verified, they can be utilized for sizing different machine ratings within a given frame size.

Design Description   Table 4.3 lists the calculated design parameters and component dimensions for the example 10-MW generator. An average current density of 100 A/mm^2 is assumed overall in the field winding region. Each field coil's cross section is assumed to span from half the pole angle (α) of 30^{oE} to an angle (β) of 86^{oE}. These are typical values used in designing such generators. In practice, the arch-type field winding cross section can be built by stacking individual pancakes or directly winding poles on a former. The inside and outside radii of the field winding region are 350 and 390 mm, respectively. The active length of the machine is 1610 mm. An HTS 1G wire (4.5-mm width × 0.3-mm

Table 4.3   Calculated design parameters and component dimensions

Parameters	Units	Values
Field Winding Assumptions and Design		
Overall current density in field region	A/mm^2	100
Half of the pole angle (α)	deg. Elect.	30
Field winding span angle (β)	deg. Elect.	86
Inside radius of field winding (R_1)	mm	350
Outside radius of field winding (R_2)	mm	390
Active length of machine (l)	mm	1610
HTS wire type	—	Bi-2223
HTS wire thickness	mm	0.3
HTS wire width	mm	4.5
HTS coil fill factor	—	0.72
Operating temperature of HTS coils	K	35
HTS coil N-value	—	12
Expected power loss in HTS coils	W	20
Field winding current	A	187.5
Turns/pole	—	2572
Mean-turn length of a coil	mm	4004
Total length of wire for all coils	km	61.8
Electromagnetic(EM) Shield (Damper) Design		
Inside radius of EM shield	mm	430
Radial thickness of EM shield	mm	40
Air gap length between shield and armature	mm	140
EM shield materials (SS inside Cu shell)	—	Copper/SS
Thickness of copper in EM shield	mm	24
Thickness of stainless steel in EM shield	mm	16
Armature Winding Assumptions and Design		
Overall current density in armature region	A/mm^2	1
Inside radius of armature winding (R_3)	mm	570
Outside radius of armature winding (R_4)	mm	647
Slots/pole/phase	—	8
Turns/coil	—	4
Turns in a coil height	—	2
Coils/slot	—	2
Armature conductor fill factor (λ)	—	0.69
Coil pitch (ξ)	—	5/6
Number of circuit	—	2
Cooling type	—	Oil
Electric stress in insulation	kV/mm	5
Total insulation thickness on coils	mm	1.593
Insulation thickness on turns	mm	0.319
Insulation thickness of ground plane	mm	1.275
Litz wire strand diameter	mm	0.5
Cooling water temperature	°C	35

Table 4.3 *Continued*

Parameters	Units	Values
Armature Winding Assumptions and Design		
Temperature rise in heat exchanger	°C	5
Temperature rise of coolant through coils	°C	15
Number slots	—	144
Armature turns/phase	—	192
Winding factor, k_w		0.923
Mean-turn length of a coil	mm	4773
Slot width	mm	15.3
Tooth width	mm	9.6
Coil height	mm	38.5
Conductor width	mm	12.1
Conductor height	mm	8.35
Copper fraction in a conductor	mm²	69.9
Total length of conductor in whole armature	m	2749
Peak temperature of copper	°C	55
Armature resistance/phase	mΩ	0.4
Inside radius of iron yoke	mm	652
Outside radius of iron yoke	mm	717
Average flux density in iron yoke	tesla	1.6

thickness) is assumed. The field winding is operated at 35 K in order to minimize the wire cost at a moderate refrigeration system cost. The operating field current is 187.5 A and is constrained by the local magnetic field and temperature experienced by the HTS wire. All the poles of the field winding require 61.8 km of the HTS wire. The field winding is enclosed in a cryostat that isolates it thermally from the room-temperature environment. During operation any asynchronous harmonic fields are likely to induce AC losses in the HTS windings. These losses must be removed with a refrigerator, which is costly. To minimize these losses, an electromagnetic (EM) conductive shield is placed outside the field winding at a radius larger than R_2. In principle, the following three locations are possible for such an EM shield:

1. Next to the field winding at a radius slightly larger than R_2. The operating temperature of the shield is about the same or slightly higher than the field winding.

2. At an intermediate radius between R_2 and radius of warm wall of the cryostat. The EM shield in this location could be operated at a temperature higher than the field winding temperature. Losses in the EM shield are thus at a higher temperature and so can removed more economically.

3. Outside of the room-temperature cryostat wall. The EM shield in this location experiences losses at room temperature that are easily removed by the air in the air gap.

The location 3 is selected for this design because this shield is the most robust and has minimal mechanical and cooling risks. The inside and outside radii of the EM shield are 430 and 470 mm, respectively. The total radial thickness of the EM shield is 40 mm, which consists of 24-mm-thick copper shell supported on the inside with a 16-mm thick stainless steel shell. The copper shell attenuates any asynchronous fields, and the stainless steel shell mechanically supports copper shell during short-circuit faults.

The stator houses a three-phase armature air-gap type double-layer winding. In the simplest terms, the air-gap winding is similar to a conventional winding housed in magnetic iron-core slots with one difference—the iron core is replaced with a nonmagnetic, nonconductive material such as G-10 or an equivalent epoxy structure. This nonconductive structure must bear all mechanical forces and allow efficient cooling of the copper armature winding. The double-layer armature winding is contained in 144 slots (with 8 slots/pole/phase). Each slot accommodates two coils and each coil has four turns. The two coils are housed in a slot. Coils are cooled with oil flowing between a coil side and a slot wall. The coil insulation thickness is determined on the basis of an electric stress of 2.5 kV/mm. The armature winding occupies an annular space between radii R_3 and R_4 of 570 and 647 mm, respectively. Since the air-gap winding is not magnetically shielded by iron teeth, the armature conductors are made from multiple 0.5-mm diameter strands configured as a Litz wire. An iron yoke encloses armature winding outside of radius R_4. The iron yoke is designed for an average magnetic field of 1.6 tesla and has inside and outside radii of 652 and 717 mm, respectively.

Table 4.4 provides performance parameters for the generator. Due to absence of magnetic iron in the active air-gap region, the mutual coupling between the rotor and stator windings is usually quite small. This leads to a lower synchronous reactance. However, the field winding time constant is very large because the field winding resistance is normally small. The EM shield (damper) time constants are comparable to those of conventional machines. The impact of these parameters on machine performance is discussed later. Because of the low synchronous reactance, the load angle at the rated load is also small. A small load angle is usually allows for better transient stability of the machine while operating on an electric grid.

Table 4.4   Performance parameters

Parameters	Units	Values
Stator line voltage	kV	13.8
Stator current	A	511
Base impedance	Ω	15.6
Armature resistance	mΩ	1.6
Field current	A	187.5
Field circuit resistance	mΩ	0.569
EM shield (damper) resistance	μΩ	9.24
Mutual reactance between armature and field	Ω	67.25
d-Axis synchronous reactance	pu	0.2
d-Axis transient reactance	pu	0.19
d-Axis subtransient reactance	pu	0.140
q-Axis synchronous reactance	pu	0.2
q-Axis subtransient reactance	pu	0.141
Armature short-circuit time constant	s	14.3
d-Axis transient short-circuit time constant	s	152,400
d-Axis subtransient short-circuit time constant	ms	18
d-Axis transient open-circuit time constant	s	162,500
d-Axis subtransient open-circuit time constant	ms	24
q-Axis subtransient short-circuit time constant	ms	24
q-Axis subtransient open-circuit time constant	ms	34
Load angle at rated load	deg	8.8

Operation   A superconducting machine has many advantages and a few challenges for its operation on an electric grid. The absence of iron in most of the magnetic circuit causes these machines to have a very low synchronous reactance (typically 0.2–0.5 pu), which provides among other benefits a much larger dynamic stability limit within its MVA rating. A rapid feedback control system is not often required for stabilization of these machines. They also have superior damping during small oscillations and require no field forcing for damping these oscillations. Generally, superconducting machines are more robust than conventional machines during transient system faults. However, transient and subtransient reactances are similar to those of conventional machines.

The lower synchronous reactance permits operation of these machines at lower load angles than conventional machines. Additional benefits of a superconducting machine are summarized below:

- Supply reactive power (MVARs) up to its full MVA rating (both leading and lagging).
- Compact and lighter.

- Virtually no harmonics in the terminal voltage.
- Improved rotor life because of elimination of thermal load cycling due to field winding current changes.
- Higher efficiency, even under partial load conditions leading to significant operating cost savings.
- Lower vibrations and noise.

When a synchronous machine operates at lagging zero power factor, armature current (I_a) produces a magnetic field that directly opposes field generated by the field winding on the rotor. In other words, field winding must generate extra voltage (V_{ar}) to overcome armature reaction. Synchronous reactance (x_d), a fictitious quantity, is equal V_{ar}/I_a. In a conventional iron core machine, x_d is strongly influenced by the field and armature currents due to the saturation of iron. Because of the absence of iron in a superconducting machine, x_d remains essentially constant over the whole operating range of such a synchronous machine.

Figure 4.3 shows relationship among generator quantities (terminal voltage, armature current at a lagging power factor, armature resistance, synchronous reactance and internal voltage generated by field current). When the example generator (described in Table 4.4) operates at a power factor of 0.85 lagging, its load angle is only 8.8° and the internally generated voltage is 1.12 pu. Both quantities are very small compared to those experienced in a conventional iron-core machine. Because of the low x_d this machine has an excellent voltage regulation capability. Such a machine may not need an automatic voltage regulator for its stable operation on an electric grid.

A superconducting machine can operate over its whole operating range without concern for static or dynamic stability. Figure 4.9 shows V-curves for the example machine for 5 load points from no-load to full rated load. In this figure the normal excitation (i.e., field current) is that needed to generate a rated voltage at the machine terminals under an open-circuit condition (i.e., armature current is zero). At no-load this machine can deliver its rated current (with a leading zero power factor) with only 80% of its nominal excitation. At this point the generator acts as an inductor and delivers inductive power to a grid equal to its MVA rating. This mode is useful during light load conditions (at nights) when underground cables present a capacitive load to a generator. This generator will also deliver its rated current (with a lagging zero power factor) at only 120% of its nominal excitation. At this point the generator acts as a condenser and absorbs inductive load from the grid. Other curves in the figure show amount of reactive power this generator can deliver at various levels of active load. The generator

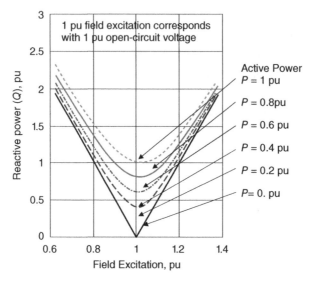

Figure 4.9 *V* curves for various powers (*P*)

remains stable over the whole range of the V-curve; it is only limited by the armature's current. A superconducting generator is an ideal synchronous condenser. An 8-MVAR unit was built and tested [11] for a year in an electric grid during 2006.

Operation with Variable Speed Drive (VSD)   Many applications of large synchronous motors require that they operate with a VSD that supplies the motor with power of varying voltage and frequency. A superconducting motor with a room-temperature EM shield is an ideal device for such applications. Because of the large air gap between the armature and EM shield, the space and time harmonics from the armature winding have a minimal impact on the EM shield and the superconducting field winding. A prototype 5-MW, 230-RPM superconducting synchronous motor for ship propulsion application was successfully tested [10]. A 36.5-MW, 120-RPM full-size motor was also successfully factory tested with a VSD [12].

4.5 MANUFACTURING ISSUES

The manufacture of a superconducting machine is quite similar to that of a conventional machine except for the following components:

- Superconducting field winding and its cooling system.
- Torque transfer from the superconducting field winding to a warm shaft.
- Stator winding.

Manufacturing issues relating to these areas are discussed below.

4.5.1 Superconducting Field Winding and Its Cooling Systems

Almost all superconducting machines built through 2010 utilized 1G HTS (BSCCO-2223) material because it was the only material then commercially available. Its electrical, mechanical, and thermal capabilities allowed introduction into electric machinery applications. Although, 2G HTS materials based on thin films of YBCO were introduced commercially in 2007, their performance and cost is still not in par with the 1G HTS material.

Because of the flat geometry for 1G and 2G wires, pancake- or layer-type coils appear to be the best building block for constructing a field winding. The assembly and support of HTS coils in a field winding configuration is challenging. Various manufacturers have developed intuitive solutions for the support and cooling of these windings. In most cases these windings are cooled by conduction; that is, the winding heat load is conducted to a cold member, which in turn is cooled with a suitable cryogen that interfaces between rotating field winding and stationary refrigerators.

In the pancake construction technique, each pancake coil is individually constructed and epoxy impregnated. Necessary number of coils is then assembled to create a field winding pole as shown in Figure 4.10. Layer winding approach for constructing a field winding is shown in Figure 4.11 where HTS wire is directly wound into slots in the rotor. Once all the poles are assembled on the rotor, they are enclosed in a cryostat to maintain the cryogenic environment for their operation. When operated in magnetic fields above 2T, 1G wire was limited to temperatures lower than 40 K. Although the 2G materials offer a potential for further cost reduction, mechanically stronger wires, and operating temperatures closer to 77 K, these features are yet to be demonstrated in electrical rotating machines. Both 1G and 2G conductors in their standard form carry only about 100 A, which is quite small for constructing field winding for a large motor or generator. Many turns are required, so this increases coil inductance and makes the task of changing field current rapidly nearly impossible. It is also not possible to

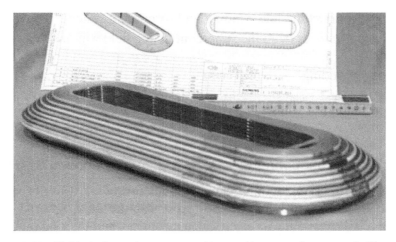

Figure 4.10   Field winding pole constructed by stacking pancake-type coils (Courtesy of American Superconductor Corporation)

Figure 4.11   Rotor field winding wound directly in rotor slots (Courtesy of Siemens)

employ industry standard exciters and voltage regulators. Currently continuously transposed cable (CTC) conductors are being developed wherein a 2G conductor tape is cut in a zigzag fashion to create the strand of desired width. These tapes are then assembled to create a "Roebel" type of cable that can carry large current (2000–3000 A). Field winding produced with such cables will have very few turns. This would permit use of industry standard exciters and standard voltage regulators. Future machines are expected to utilize such CTC

Figure 4.12   Cooling system for a 5-MW, 230-RPM superconducting motor (Courtesy of American Superconductor Corporation)

cables. It might also be possible to build armature AC coils with the CTC cable.

Most of the HTS machines prototyped so far have employed the G-M cryocooler refrigerators discussed in Chapter 3. A single-stage G-M cooler can supply 100 W of cooling power at 30 K. Cooling systems for larger machines can be built by using the G-M cryocooler as the building block. It employs gaseous helium as a working fluid to transport heat from HTS windings to the cryocoolers. A cooling system built by AMSC for the 5-MW, 230-RPM motor is shown in Figure 4.12.

Field windings are cooled by conduction to a cooling media supplied from a stationary refrigerator. Two popular choices for cooling media have been liquid neon and gaseous helium. The coolant is passed on to the rotor through a rotating coupling, which allows the inflow and outflow of coolant. The cooling transfer systems have been successfully built and operated by many different companies: first as part of LTS projects during the 1970s and 1980s and more recently as part of HTS projects. In all cases the technical challenges for the design of this component were solved.

4.5.2 Torque Transfer from Cold Field Winding to Warm Shaft

A mechanical torque is directly experienced by the field winding operating in a cryogenic environment. The torque must be transferred from the cryogenic environment to a warm shaft at room temperature with minimal thermal conduction from the warm to cryogenic environment. Options and issues relating to the torque transfer are discussed in Section 4.3.2.

4.5.3 Stator Winding

In the manufacture of the stator, two design options are available:

- Conventional stator utilizing magnetic teeth for guidance of the flux ("retrofit").
- Air-core winding without magnetic teeth.

In some application areas like large power generation there may be a strong desire to keep the machine as compatible with conventional designs as possible. This favors the conventional stator concept with magnetic teeth. In case of the retrofit solution, the air gap could be increased to limit the reaction from the stator on the superconducting rotor winding. An insertion of a HTS rotor in a conventional stator could also allow a small upgrade in the capabilities of an existing machine. This strategy is of benefit both for the customer and the manufacturer's service and retail business. It helps overcome the barriers to the introduction of new machines consisting of novel HTS rotors and challenging and expensive stators.

The use of a high-performance air-core stator winding calls for a more compact design, smaller by a factor of 4 (or better) compared to conventional machines. The small synchronous reactance of an HTS machine provides better stability, high overload capacity, and improved reactive power compensation capability. Such concepts have already been developed and successfully tested within LTS generator projects and HTS motor programs.

Although manufactured from "conventional" copper, the stator air core winding represents a big challenge. The high power density in HTS rotating machines is achieved by employing high-performance mechanical and electrical components in the stator windings. Forced-flow cooling systems are usually required for high power density machines. An obvious solution is water-cooling, as is widely used in high rating machines. This approach was employed by GE [6] and Super-GM

[13] in construction of their air-core stator windings. An alternate approach may be to employ fluids to function as both heat transfer and electrical insulation. Example fluids are "Midel" and "Silicone Oil." Heat pipes or thermal siphons could also be employed to address extreme design challenges in this area.

4.6 SIMULATION

Superconducting motors and generators need an analytical model for simulation in an electric grid or for control with power electronics. The simulation model should be similar to that currently in use by commercially available codes such as PSCAD, EMTP, PSSE, and MATLAB. Most of these codes represent a synchronous machine with Park's two-axis (d-q) equations [14]. It is possible to model a superconducting machine with d-q axis equations where mutual inductances among various coils are not equal. However, almost all commercial codes assume a common mutual inductance among all coils on each axis (d and q-axes). These codes work with equivalent circuit and per-unit system similar to that described in Section 4.2.3. The author has studied transient behavior of superconducting machines with and without assumption of equal mutual inductance among various coils on each axis and has observed that results with the two methods are within ±5%. For most practical applications, this error is small, and it is preferable to employ commercial codes that come with built-in tools to interface with an electric grid or power electronics and with all necessary graphic capability.

4.7 GENERATORS

Superconductivity has been a hope and vision for large power generation since the early 1960s and 1970s. At that time the development of nuclear power generation pushed the generator ratings to the physical size limits, and the superconductivity technology was acknowledged as an opportunity to further increase power densities and ratings. The additional major technology benefits were improved efficiency, increased reactive power capability, and reduced synchronous reactance with a positive impact on power system stability. A number of generators using LTS NbTi wire were successfully demonstrated all over the world between 1970 and 1990. They were not considered economically attractive because of the complexity of the cryogenic system and the poor stability of the LTS windings.

However, the invention of HTS in the mid-1980s provided the incentive to look at rotating machines again. HTS coils can operate at substantially higher temperatures than those made of LTS materials, and can utilize relatively simpler, less costly, and more efficient refrigeration systems. These factors make HTS wire technically suitable and economically feasible for use in the development and commercialization of motor and generator applications at power ratings much lower than could be considered with LTS wire. Two types of generators have been prototyped in recent years:

- High-speed generators (>10,000 RPM)
- Low-speed–AMSC (<3600 RPM)

4.7.1   High-Speed Generators

Military and commercial applications need 1 to 5 MW capability in a portable high power density electric power generation package. One approach is to use a high-speed generator directly coupled to a high-speed gas turbine, generate power at high frequency, and convert this power at desired frequency using solid-state power conversion techniques. Superconducting technology offers the highest power density generators, but several engineering challenges remain in making a reliable light-weight superconducting machine. To address this need, a rugged, high-speed, multi-megawatt, HTS generator was developed [15] by GE for the Air Force Research Lab (AFRL). The generator was load tested up to 1.3 MW at GE's high-speed machine test bed. The generator is based on the homopolar inductor alternator (HIA) [16] topology to obtain power densities greater than 4 kW/lb in a robust construction suited for high-speed applications.

The generator comprises a stationary HTS field excitation coil, a solid rotor forging, and an advanced but conventional stator, as shown in Figure 4.13. The armature consists of liquid-cooled air-gap windings placed within an advanced iron yoke with laminations oriented in three dimensions to carry flux from one end of the machine to the other. The stationary HTS field coil is placed around the ferromagnetic rotor forging and between two sets of salient poles that are offset circumferentially by one pole-pitch at either end. The HTS field coil provides MMF capability much higher than a traditional copper coil, enabling an "air-gap" armature winding with high flux density in the gap. Key features of this generator are listed below:

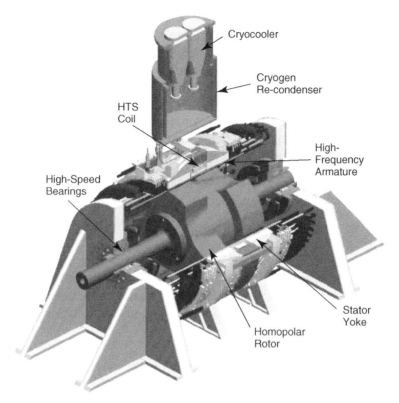

Figure 4.13 Schematic of homopolar inductor alternator with HTS field winding (Courtesy of General Electric)

- The stationary superconducting field coil does not experience the large centrifugal forces that a rotating coil experiences. The thermal insulation between the coil and ambient is also improved because of lack of centrifugal loads and reduced requirements on the coil support.
- The cryostat of the coil is stationary. There is no need for a transfer coupling to introduce a cooling medium into the rotating cooling circuit.
- There is no need for a "slip-ring" and brush assembly to transfer current to the coil from a stationary exciter.

The armature winding employing conventional coils in conventional iron-core slots could be used in such machines. However, GE chose to employ air-gap winding to achieve higher power density. The air-gap armature winding design utilized coils wound with compacted Litz copper wire turns and cooled directly by ordinary water or a dielectric fluid through alumina ceramic cooling tubes. Each coil was wet wound in a precision mold with thermally conductive epoxy and cured. The coils are then assembled and adhesively bonded to the ceramic cooling tubes and G10 cylindrical inner and outer shells using thermally conductive epoxy. The G10 shells on the ID and OD of the armature coil sides served as ground wall insulation. The armature winding assembly was bonded to the stator yoke to form a rigid structure capable of withstanding fault torques, vibration, and shock loads.

4.7.2   Medium-Speed Generators

Several low-speed machines were attempted by several companies during the decade starting the year 2000. Three most notable examples are the General Electric 100-MW generator, the AMSC 8-MVAR synchronous condenser, and the Siemens 4-MW generator. These are discussed below. Very slow speed synchronous machines are also of interest for ship propulsion motors, hydroelectric generators, and wind turbine generators.

GE 100-MW Generator Project   General Electric attempted development of a 100-MW, 3600-RPM, two-pole generator by combining a superconducting rotor winding with a conventional iron-core stator. The major benefit of the adaptation of HTS generators into power plants is increased system efficiency. Generators lose power in the field windings and the armature bars. When superconducting wire is used for the field windings, the losses in the rotor can be practically eliminated. Other losses can also be reduced because of increases in power density and the reduction in the required cooling capacity. HTS generators produce electric power with lower losses than their conventional equivalents. Even small-efficiency improvements can produce big dollar savings. A half of 1% improvement in generation efficiency provides a utility or independent power producer with additional capacity to sell energy with a value of nearly $200,000 a year per 100-MW generator, at electricity prices of five cents per kWh and 8000 operating hours per year. The GE 100-MW generator utilized a conventional stator core and frame with its magnetic structures and adds an HTS rotor that contains an iron core. This approach completely eliminates any risk associated with the design of the stator and provides a magnetic structure in the rotor to enhance torque transmission. It offers immediate efficiency benefits, compatibility with the turbine drive train, and the ability to retrofit HTS rotors into existing generators. Alternatively,

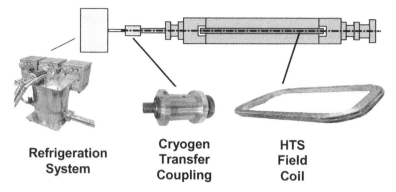

**Refrigeration
System**

**Cryogen
Transfer
Coupling**

**HTS
Field
Coil**

Figure 4.14   General Electric 100-MW superconducting generator rotor concept
(Courtesy of General Electric)

"air-core" designs can eliminate much of the structural and magnetic steel, obtaining a generator that can be smaller and lighter than an equivalent conventional generator. In applications where size or weight reduction is an advantage, such as ships or locomotives, superconducting generators could increase generating capacity without using additional space. Construction, shipping, and installation may be simplified and perhaps less costly as a result of the smaller dimensions and lighter weight. However, the challenges to producing air-core HTS generators include the transmission of torque within the generator and the potential for amplification of fault torques on the turbine-generator drive train.

Rotor configuration adapted for the 100-MW generator [7] is shown in Figure 4.14. This program was cancelled just after completing rotor design because HTS (1G) wire and refrigeration system costs were too high [17] to make the generator economically attractive. Future technologies (2G wire and pulse-tube refrigerators) were not sufficiently developed in 2004 to make an objective decision.

AMSC Synchronous Condensers   Superconducting generators have inherently a low synchronous reactance that enables such a generator to operate at any power factor (leading or lagging) while connected to an electric grid. AMSC built 8- and 12-MVAR prototype synchronous condensers for electric grid applications. An 8-MVAR dynamic synchronous condenser [18] (DSC) was operated on Tennessee Valley Authority (TVA) grid to supply reactive power to an arc furnace in order to limit the flicker. Although the prototype met all its goals, the program did not succeed for economic considerations. In other words, this machine was too expensive compared with similar size electronic solutions (STATCOM) due to the high cost of HTS materials and refrigeration systems.

The machine had a high temperature superconductor (HTS) field winding and a conventional stator winding. It had a small footprint and was readily transportable in a trailerized format, making it easily placed in substations. This HTS DSC enhances system utilization through reactive compensation by dynamically injecting leading or lagging VARs as needed. At the test site (next to an arc furnace) the testing was intended to show not only that the HTS DSC could be a preferred option for handling arc furnace flicker but also that the device can handle hundreds of thousands of transmission transients daily. The machine continuously absorbed transient disturbances with very high negative (>30%) and zero (>15%) sequence current components. The rotor with a continuous damper winding in the form of a copper shell absorbed heating created by negative and zero sequence currents very effectively with an insignificant temperature rise. The stator winding also withstood these currents while staying within the peak allowable temperature.

This HTS machine used less than half of the energy of a conventional synchronous condenser and about the same amount of energy as a modern flexible AC transmission system (FACTS) device consumes. The machine was first synchronized with the grid in October 2004 and was operated for a year per terms of the contract with TVA. The 8-MVAR synchronous condenser is shown conceptually in Figure 4.15.

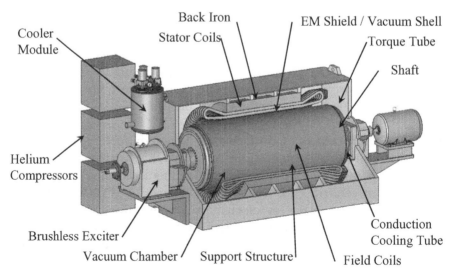

Figure 4.15 Synchronous condenser concept (Courtesy of American Superconductor Corporation)

Figure 4.16   An 8-MVAR synchronous condenser as tested (Courtesy of American Superconductor Corporation)

The figure shows various components of the machine. The actual machine as tested at the TVA site is shown in Figure 4.16. The prototype version was rated 8MVAR and production version was rated 12 MVAR. The line voltage was 13.8 kV and was designed with four-poles for operation on a 60-Hz grid.

Siemens Synchronous Generators   Siemens demonstrated two machines [19] between 1999 and 2006. From 1999 to 2001 a 400-kW model machine was constructed and manufactured in order to demonstrate the feasibility of the conceptual design for a superconducting synchronous machine with HTS. This machine, shown on the left in Figure 4.17, employed a rotating cryostat serving as a vacuum containment, which accommodated the HTS winding mounted on a four-pole iron core. The stator contains an armature winding made of copper and designed as an air-gap winding. Magnetic iron teeth were replaced by a nonmetallic support structure. The thermal conduction losses of the cryostat down to the HTS coils' operational temperature of 25 K were kept low with a specially designed torque transmission tubes. The cooling system consisted of a commercial off-the-shelf Gifford–McMahon cryocooler combined with a thermosyphon cooling process

Figure 4.17 HTS machines built by Siemens (Courtesy of Siemens)

using liquid neon as the coolant. During an intensive two-year test phase it was shown that all components, like rotating cryostat or HTS windings, could be designed and manufactured in a way that they are able to operate in a rotating machine reliably. The machine was tested in generator operation mode as well as in motor operation.

Starting in 2002, Siemens designed and manufactured its second HTS synchronous machine shown on the right in Figure 4.17. This machine was designed as a generator for marine vehicle applications rated 4 MVA at 60 Hz and 3600 RPM. Consequently it was designed and tested according to the particular standards for marine vehicle applications of the "Germanischer Lloyd." Figure 4.17 shows the machine in the company's test facility in Nuremberg, Germany.

The rotor consists of HTS pancake coils manufactured from Bi 2223-tape. The cooling system consisted three cold heads installed as shown on the right side in Figure 4.17. Each of these cold heads offers a cooling power of about 45 W at the rated operating temperature of about 25 K.

For normal operation, two cold heads are sufficient. The system is built in such a way that at least one cold head can be completely removed from the system for maintenance without affecting the operation of the machine.

Converteam Hydroelectric Generator Converteam has also completed construction of a 1.7-MW, 28-pole, 214 RPM hydroelectric generator [20] employing HTS field windings and a conventional stator. The generator is to be installed at a hydropower station in Hirschaid, Germany, and is intended to be a technology demonstrator for the practical application of superconducting technology for sustainable and renewable power generation. The generator is intended to replace and uprate an existing conventional generator and will be connected directly to the German grid. The HTS field winding uses Bi-2223 tape conductor cooled to about 30 K by high-pressure helium gas, which is transferred from the static cryocoolers to the rotor via a bespoke rotating coupling. The coils are insulated with multi-layer insulation and positioned over laminated iron rotor poles at room temperature. The rotor is enclosed within a vacuum chamber, and the complete assembly rotates at 214 RPM.

4.8   MOTORS

Industrial motors operating up to 5000-hp, 1800 RPM have been successfully demonstrated, but their introduction into the industrial market is awaiting the availability of low-cost 2G (YBCO) conductors. Whereas very slow speed synchronous machines for use in ship propulsion and wind turbine represent a strong need, their economic viability is strongly tied to the cost of 2G HTS materials. Two slow speed ship propulsion motors (5 MW, 230 RPM, and 36.5 MW, 120 RPM) built by AMSC have been delivered to ONR in the United States.

4.8.1   High-Speed Motors

Several superconducting machines employing HTS wire were built by a variety of manufacturers between 1995 and 2007. Rockwell Automation were first to start development of HTS machines [21] in the United States. Figure 4.18 shows a 200-hp (tested in 1996) and a 1000-hp (tested in 2000), four-pole, 60-Hz prototypes built by Rockwell Automation. American Superconductor (AMSC) supplied the HTS field poles made from BSCCO-2223 and their cooling systems. The 1000-hp HTS motor was designed for operation from a VSD. It employed an air-core stator winding and HTS field winding operating at 1800 RPM. In this topology the HTS coils are located on the rotor, and the stator is comprised of a conventional copper winding. The HTS coils are mounted on, and the conduction cooled through the support structure on which they are mounted. The stator winding was water-cooled with embedded cooling tubes. This approach enhances the cooling effectiveness and offers the potential for running the HTS motor at a higher rating.

200hp, 1800-RPM Motor   1000hp, 1800-RPM Motor

Figure 4.18   Rockwell Automation's 200-hp (left) and 1000-hp (right) synchronous motors employing HTS field windings (Courtesy of Baldor/Rockwell Automation)

Figure 4.19   AMSC's 5000-hp, 1800-RPM synchronous motors employing HTS field windings tested in 2001 (Courtesy of American Superconductor Corporation)

AMSC designed, built, and tested a 5000-hp; four-pole, 1800-RPM machine shown in Figure 4.19 in 2001. The motor employed closed cycle Neon heat pipe concept [22], which provided very satisfactory performance. This 5000-hp motor accomplished all its objectives: validated the design tools, confirmed the design concepts for the HTS rotor and its cooling system, and presented a novel single-layer air-gap winding concept [23].

4.8.2   Low-Speed Motors

As was stated earlier, the slow-speed synchronous machines are of interest for ship propulsion motors, hydroelectric generators, and wind turbine generators. Such machines operate at speeds ranging from 10 to 200 RPM. Because of their low speed, such machines have the characteristics of very high torque and large size and weight. Their superconducting field windings can generate a large magnetic field in the armature region, which helps in reducing the size and weight of such machines.

In late 1990s, the US Navy decided that its future combatant ships would be all electric; that is, electric power would be used for propulsion motors as well as for other loads. New technologies and techniques were considered essential to managing the generation and utilization of such electric power on a ship. Moreover lighter and more compact subsystems capable of fitting on ships that have very constricted space were sought. HTS propulsion motors were the ideal candidate for such ships. In addition these motors are compatible with standard variable speed drives (VSDs) and they meet both US Navy and commercial electric ship requirements, reducing the drive development effort and recurring costs. The range of benefits and advantages that HTS ship propulsion motors offer are listed below:

- Up to four times higher torque density than alternative technologies, HTS machines are more compact and lighter in weight. The size and weight benefits make HTS machines easier to transport and install, as well as allowing for arrangement flexibility in the ship.
- Absence of iron stator teeth reduces the structure-borne noise.
- High efficiency from full to low speed can boost the key mission parameters for ships, such as fuel economy, sustained speed, and mission range.
- Isothermal field windings are well suited for repeated load changes.

The significant advantages to an HTS motor propulsion system, when compared to conventional technology, are provided by the absence of iron stator teeth. A large air gap is formed that allows designers to maximize the power density while independently tuning the machine parameters, such as synchronous and subtransient reactances, to meet the following system needs:

- Operation at unity power factor to lower the electric drive rating and cost.

- Lower capacitance to ground to minimize the impact of the drive switching frequency on the motor.
- Manageable fault current (achieved by operating the HTS motor at higher voltages, up to 10 kV) to reduce the breaker requirements.
- Accommodation of larger harmonic currents than conventional machines due to the attenuation of harmonic fields in the larger air gap and the capability to withstand heating in the conductive EM shield on the rotor. This feature also reduces the need for harmonic filters and their associated weight and volume.

AMSC built two prototype motors under contract with the Office of Naval Research: 5-MW, 230-RPM motor [24] and 36.5-MW, 120-RPM motor [12].

At the heart of this power-dense technology lies a rotor constructed with superconductor wire that permits the generation of much higher magnetic fields than is practical with conventional copper windings, or possible with present permanent magnet technology. To best take advantage of this very high rotor flux, the teeth holding the stator windings are nonmagnetic. The stator is liquid cooled with a dielectric insulating oil. The 36.5-MW motor represents a 14:1 increase in torque over the 230-RPM, 5-MW HTS motor previously built and tested. Both motors employed 1G (BSCCO-2223) field windings cooled with G-M cryocoolers.

The 5 MW motor coupled to two 2.5-MW induction motors is shown in Figure 4.20 at the CAPS facility. CAPS operated the motor at full load (5 MW) and at full speed (230 rpm) for a total of 21 hours, and confirmed that the motor achieved steady-state temperature both in the rotor and in the stator. The actual temperatures measured correlated closely to designed values for the machine. This load testing demonstrated that the HTS motor meets its specified performance and power rating under the stresses of operating conditions.

The objective of the HTS 36.5-MW motor was the development of a full-scale, high power density, lightweight, advanced propulsion motor suitable for future naval applications. As built, the 36.5-MW, 120-RPM HTS motor is shown in Figure 4.21. The 36.5-MW motor is a scaled version of the 5-MW HTS motor. It employs many of the technologies developed for the 5-MW motor, including the HTS field winding, a high current density liquid-cooled stator, and the refrigeration, mechanical support, and excitation systems. With an output shaft speed of 120 RPM, it generates over 2.9 million Newton-meters (2.2

Figure 4.20 AMSC's 5-MW motor coupled to two 2.5-MW induction motors (Courtesy of American Superconductor Corporation)

Figure 4.21 AMSC's 36.5-MW motor undergoing factory testing (Courtesy of American Superconductor Corporation)

million ft-lbs) of torque. The design drivers include low system weight; other drivers are improved efficiency and shock capability. The US Navy performed the full-load testing of the 36.5-MW HTS motor system at a the Navy-operated land-based test site in Philadelphia in late 2008.

4.9 SUMMARY

Many developers have amply demonstrated the technical feasibility of HTS rotating machines at both high speed and low speed, but their economic feasibility is contingent on the availability of low-cost super-conductors and reliable and affordable cooling systems. The cost of the wire and the cooling system must be low to make HTS machines competitive with conventional copper and permanent magnet machines. The fact that the HTS machines could be smaller, lighter, and more efficient than the conventional copper and permanent magnet machines has been demonstrated by the 36.5-MW, 120-RPM ship propulsion motor built for the US Navy. Even larger machines are possible with this technology. Direct-drive 10-MW wind turbine generators have proved feasible for the offshore applications. Once the HTS technology becomes affordable, the possible applications could include motors and generators for central power stations, wind turbine generators, ship propulsion motors, industrial motors, and synchronous condensers, for example.

REFERENCES

1. H. H. Woodson, Z. J. J. Stekly, and E. Halas, "A study of Alternators with Superconducting Field Windings: I. Analysis," *IEEE Trans. Power Appara. Syst.* 85: 264–274, 1966.

2. J. L. Kirtley, "Basic Formulas for Air Core Synchronous Machines," Paper 71 CP 155-PWR, Winter Power Meeting, IEEE, New York, 1971.

3. J. R. Bumpy, *Superconducting Rotating Electrical Machines*, Clarendon Press, Oxford, 1983.

4. B. Adkins, "The General Theory of Electrical Machines," Chapman and Hall, 1964.

5. S. S. Kalsi, and P. M. Winn, "Thermally Conductive Stator Support Structure," US Patent 7,423,356 and 7,211,919.

6. Superconducting Generator Design, EPRI EL-663, Project 429-2, Final Report, March 1978.

7. S. S. Kalsi, K. Weeber, H. Takesue, C. Lewis, H.-W. Neumueller, and R. D. Blaugher, "Development Status of Rotating Machines Employing Superconducting Field Windings," *Proc. IEEE* 92(10): 1688–1704, 2004.

8. R. McFee, "Optimum Input Leads for Cryogenic Apparatus," *Rev. Scientific Instr.*, 10(2): 1959.

9. S. S. Kalsi, D. Madura, and M. Ingram, "Superconducting Synchronous Condenser for Reactive Power Support in an Electric Grid," *IEEE Trans. Appl. Superconductivity* 15(2, Pt. 2): 2146–2149, 2005.

10. G. Snitchler, B. Gamble, and S. S. Kalsi, "The Performance of a 5 MW High Temperature Superconductor Ship Propulsion Motor," *IEEE Trans. Appl. Superconductivity* 15(2, Pt. 2): 2206–2209, 2005.

11. S. S. Kalsi, D. Madura, T. MacDonald, M. Ingram, and I. Grant, "Operating Experience of Superconductor Dynamic Synchronous Condenser," *IEEE Power & Energy Society Transmission & Distribution* 899–902, 2006.

12. J. Buck, B. Hartman, R. Ricket, B. Gamble, T. MacDonald, and G. Snitchler, "Factory Testing of a 36.5 MW High Temperature Superconducting Propulsion Motor," *ASNE Day Symp.*, June 25–26, 2007.

13. K. Yamaguchi, M. Takahashi, R. Shiobara, T. Taniguchi, H. Tomeoku, M. Sato, H. Sato, Y. Chida, and M. Ogihara, "70 MW Class Super-conducting Generator Test," *IEEE Trans. Appl. Superconductivity* 9(2): 1209–1212, 1999.

14. R. H. Park, "Two-Reaction Theory of Synchronous Machines," *Trans. AIEE* 48: 716, 1929.

15. K. Sivasubramaniam, T. Zhang, M. Lokhandwalla, E. T. Laskaris, J. W. Bray, B. Gerstler, M. R. Shah, and J. P. Alexander, "Development of a High Speed HTS Generator for Airborne Applications," Presented at the 2008 IEEE Applied Superconductivity Conference.

16. E Richter, "High Speed Homopolar Inductor Generator with Straight Winding Construction," US Patent 3,737,696, 1973.

17. J. M. Fogarty, "GE Generator Status Report," 2004 US: DOE Superconductivity Peer Review.

18. S. S. Kalsi, D. Madura, M. Ross, M. Ingram, R. Belhomme, P. Bousseau, and J.-V .Roger, "Operating Experience of a Synchronous Compensator," CIGRE 2006, Paper A1-108.

19. G. Klaus, W. Nick, H-W Neumuller, G. Nerowski, and W. McCown, "Advances in the Development of Synchronous Machines with High Temperature Superconducting Field Winding at Siemens AG," 2006 IEEE Power Engineering Society General Meeting, June 18–22, 2006, Montreal, IEEE Cat. 06CH37818C.

20. R. Fair, C. Lewis, J. Eugene, and M. Ingles, "Development of an HTS Hydroelectric Power Generator for the Hirschaid Power Station," Presented at the 2009 European Conference on Applied Superconductivity.

21. D. Aized., B. B. Gamble, A. Sidi-Yekhlef, J. P. Voccio, D. I. Driscoll, B. A. Shoykhet, and B. X. Zhang, "Status of the 1000 hp HTS Motor Development," *IEEE Trans. Appl. Superconductivity* 9(2): 1999, pp 1197–1200.

22. B. B. Gamble, A. Sidi-Yekhlef, R. E. Schwall, D. L. Driscoll, and B. A. Shoykhet, "Superconductor Rotor Cooling System," US Patent 6,376,943.

23. S. S. Kalsi, "Superconducting Electric Motor," US Patent 7,453,174.

24. S. S. Kalsi, B. B. Gamble, G. Snitchler, and S. O. Ige, "The Status of HTS Ship Propulsion Motor Developments," 2006 IEEE Power Engineering Society General Meeting, June 18–22, 2006, Montreal, IEEE Cat. 06CH37818C, ISBN 1-4244-0493-2.

5

ROTATING DC HOMOPOLAR MACHINES

5.1 INTRODUCTION

DC homopolar machines are based on Faraday's disk machine demonstrated in 1831. Once low-temperature superconductors (LTS) became available in 1960s, these were the first machine prototyped [1]. Homopolar machines have characteristics of low voltage and high current. The high current is collected from rotating members with brushes. Efficient current collection with brushes is challenging, however, and is preventing commercialization of this technology. Some of the advantages of a superconducting homopolar motor are that its field coil is a simple solenoidal winding located on the stationary part and it does not experience motor torque. This simplifies design of the field winding and its cryostat. The US Navy spent many years attempting to make this technology work [2]. Large machines, in several configurations up to 5 MW, have been built using superconductor field windings.

5.2 PRINCIPLE

The homopolar machine based on Faraday's disk machine is shown in Figure 5.1. The figure shows a disk with a vertical shaft. A radially

Applications of High Temperature Superconductors to Electric Power Equipment,
by Swarn Singh Kalsi
Copyright © 2011 Institute of Electrical and Electronics Engineers

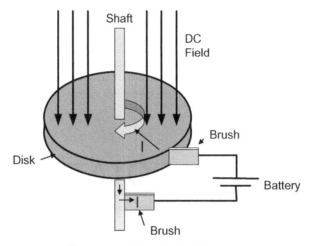

Figure 5.1   Faraday's disk motor

directed current (I) is created in the disk by applying a voltage between the shaft and rim of the disk. The interaction between the radial current and the vertical field makes the disk turn. The voltage induced in the disk is given by

$$V_i = \pi BN \left(R_o^2 - R_i^2 \right), \tag{5.1}$$

where

R_i, R_o = inner and outer radii of disk (m),
N = rotational speed (rev/sc)
B = magnetic field (T).

In a disk (with inner and outer diameters of 0.1 and 1 m, respectively) rotating at 300 RPM (= 5 revolutions/second) in a 1-T field, voltage induced between the shaft and the outer rim of the disk is 15.5 V. In a 100-kW (= 134-hp) motor, a current of 6.4 kA has to be collected from the disk. For a fixed size, the machine current increases directly proportional to its rating, since the induced voltage is fixed by applied magnetic field and rotational speed of the motor. In a practical machine, it is desirable to increase the voltage and reduce the current. The following section describes a few techniques for achieving this goal.

5.3   CONFIGURATION

One way to increase machine voltage is to connected several disks in series. This increases the number of brushes because two brushes or brush sets are needed for each disk. A more practical approach selected by developers is to employ the drum rotor configuration shown in Figure 5.2. A field, radial to the rotor's outer surface, is created with two cylindrical coils located near the ends of the drum rotor. Figure 5.3 shows the variation of radial field along the length of a rotor of a typical machine. Current is introduced at one end of the rotor and is collected on the other end using brushes. The axial distance between the brushes

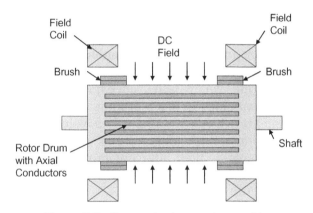

Figure 5.2   Drum rotor homopolar machine

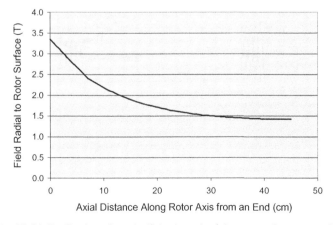

Figure 5.3   Field distribution along half the length of the rotor of an example machine

determines the active length of the machine. The voltage induced between the brushes is given by

$$V_i = 2\pi RNBl, \tag{5.2}$$

where

    R = mean radius of drum (conductors),
    N = rotational speed (rev/sc),
    B = magnetic field (T),
    l = axial (active) length (m).

In this configuration the induced voltage is controlled by increasing active length (l), but an optimum length to diameter ratio can be found for a given magnet/rotor configuration. With an active length of 1 m, an average drum radius of 0.5 m, and the same field and rotational speed as for the disc motor above, the voltage induced between brushes is 15.5 V. This is still a very low voltage. However, it is possible to employ several concentric drums where each drum experiences a common field created by the field winding [3]. In a 5-drum machine (shown in Figure 5.4), it is possible to achieve $5 \times 15.5 = 77.5$ volts by connected all drums in series. In practice, the longer drums will have higher voltage than the shorter drums as they intercept more radial flux than the

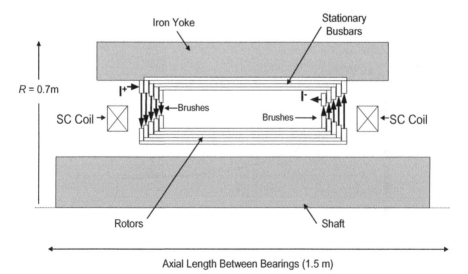

Figure 5.4   Outline of the example machine

shorter drums. Each drum requires a separate pair of brushes on each end and all brushes carry the same current. Additional drums could be added to achieve higher voltage at the expense of more brushes and complexity. The axial length occupied by the brush assemblies will result in some lost flux and lower output voltage per drum.

In the example above we assumed an effective field of 1 T, a value that could be reasonably achieved with superconducting coils. The maximum practical achievable voltage is usually on order of 300 to 400 V, which still requires a large current through many brushes. Resistive loss in the brushes and the field-induced circulating currents in the brushes can reduce efficiency with added drums and/or higher fields. As a larger current is pushed through drum conductors, their cooling becomes challenging, and active cooling of these conductors and brushes becomes necessary.

5.4   DESIGN CHALLENGES

Drum rotor type of machines are being considered for large ratings [3,4]. Major challenges of such motors are listed below:

- Superconducting field winding
- Magnetic shielding
- Drum rotor
- Brush system

Superconducting Field Winding   The design of superconducting coils is straightforward, and they can be built with LTS or HTS. Since these coils experience no torque, their structural design is greatly simplified. The only major force experienced by the superconducting coils is their mutual repelling force directed along the rotational axis of the machine. This force can be reacted with axial ties between the coils provided that they share a common cryostat. However, it is also possible to support each coil individually at the expense of an additional refrigeration load.

The choice of superconductor is straightforward for this motor. LTS NbTi coil operating at about 4.5 K could be the most the economical choice. Another LTS conductor, Nb3Sn, is also a good candidate, though it is mechanically more challenging to work with. It can also be operated at 4.5 K but has higher T_c than NbTi, allowing for higher current. The cost of the refrigeration system would be higher than that for HTS coils, but the total system cost could be lower than an HTS system due to the much lower cost of LTS and its higher performance as compared to HTS. Superconducting motors are normally operated with fixed excitation and motor speed is varied by changing armature voltage. However, it is also possible to achieve the speed control by changing the field excitation while keeping drum voltage constant. The field coils could be constructed using high-current Roebel cable made from HTS wire. Such conductors have high current capability, which reduces the number of turns and thereby substantially reduces the coil inductance. Since HTS coils can easily absorb a small heat load due to field current ramping, a homopolar machine built with such HTS coils could be controlled by varying its field current. However, resistive loss due to the drum current would increase at a reduced field for a given load. Some combination of field and drum current variations could yield optimum capital cost while enabling the most efficient operation of a machine.

Magnetic Shielding Although the back iron is not required to reduce the reluctance of the flux path in a superconducting homopolar motor, a stray field of superconducting magnets may need to be contained to reduce the impact on the surrounding equipment. This shield can be easily applied and shape optimized as part of the motor housing, which also reacts the shaft torque. The motor housing can be made from a solid magnetic iron, since the fields are only DC. The degree of magnetic shielding required can have a substantial impact on the overall motor weight.

Drum Rotor A rotor with many turns could be created by including several concentric insulated drum layers on the rotor. Each drum needs a brush set on each end for handling the current. On the stator side, a brush on the one end of a drum must be connected to the brush on the opposite end of the adjacent drum. Copper bus bars are typically employed for interconnection between the brush sets at the ends of a rotor drum. The copper bus bars are parallel to the machine's rotational axis, and each bar is on average equal to the active length of the machine. These bus bars must be supported in the machine's frame to react rotor torque. In order to achieve maximum compactness of the machine, copper in the rotor drums and bus bars must operate at highest possible current density consistent with desired efficiency and thermal management. For most compact designs, it is necessary to provide active cooling of these copper components. Active cooling of the rotor becomes complicated due to introduction and recovery of

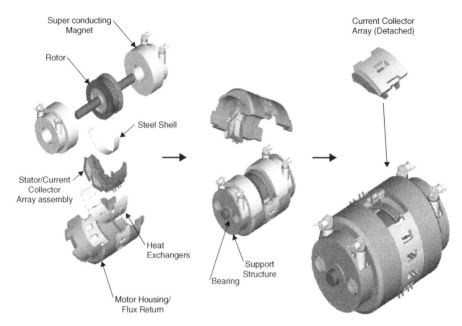

Figure 5.5 GA's 3.7-MW homopolar machine at General Atomics[†]

coolant to the individual rotor drums while maintaining electrical isolation and uniform temperature distribution.

Brush System Brushes are the biggest challenge for DC homopolar machines. The brushes are located in the stationary part of the motor (stator) and provide the electrical connection to the normally conducting, liquid-cooled rotor. Solid carbon or metal-graphite brushes were found inadequate due to their low current density capability and excessive wear. Graphite fiber brushes were used with some success in early superconducting homopolar motors.* Liquid metal brushes were also developed and applied to homopolar machines through the 1980s, but were not suitable for many applications due to material and life limitations. Currently copper fiber brushes [3] operating in a wet humidified CO_2 environment are being considered, and this provides a compromise between current-carrying capability and long-term wear. Such

*The typical combatant expects to be underway 2750 hr/yr and 6% at full speed (approx. motor speed of 12 m/s). For at typical operating profile of 2750 hr per year, the total sliding distance traveled under brushes will be approximately 65 million-miles per year.

brushes are expected to provide a five-year operational life. For a typical operating profile this equates to motor slipring travel of 6.5×10^7 m/yr [1]. A significant challenge is to control the brush losses in order to limit their maximum operating temperature. For this reason complex means for brush loading and liquid cooling of the rotor and stator are necessary.

5.5 PROTOTYPES

General Atomics (GA) has designed, fabricated, and partially tested a 3.7-MW (4960 b-hp) motor [5]. The 3.7-MW machine is shown in Figure 5.5. GA has also recently completed the preliminary design of a 36.5-MW low-speed superconducting homopolar motor (shown schematically in Figure 5.6), which is sized for ship propulsion applications. The 3.7-MW LTS motor is 3.5 feet in diameter and 6 feet long. The 36.5-MW motor would be approximately 9 feet in diameter, and 12 feet long. The reduced diameter of this machine allows for improved hydrodynamics of the propulsion system in pod applications.

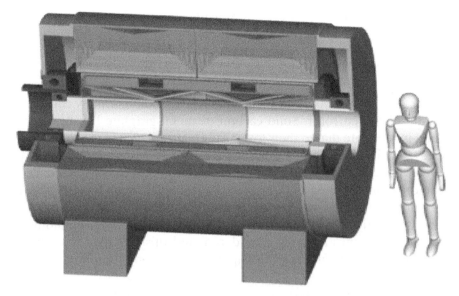

Figure 5.6   GA's 36.5-MW homopolar motor under development at General Atomics[†]

The DC homopolar motor has a number of potential advantages over HTS-AC motors. For one, homopolar motors are expected to be extremely quiet as they lack the inherent time and spatially varying forces that are a source of noise in AC electric machines. Indeed the differences between the two are characterized more by the fact that one is DC and the other AC than by the fact that one operates at 4.7 K and the other operates at about 30 K.

The attractiveness of the homopolar motor system is its relatively simple design. The superconducting DC homopolar motor system, when combined with future DC link power distribution systems, might simplify the ship's integrated power distribution and motor drive system, reduce the number of system components, improve efficiency, and lower system acquisition and operating costs. However, the DC distribution systems currently being considered are aiming at 3000- to 6000-V DC, so direct compatibility with the homopolar motor is questionable. An attractive combination that could eliminate power conversion is to mate the homopolar motor with a generator that has an intrinsic low voltage output such as fuels cells and thermoelectric generators

5.6   SUMMARY

Superconducting DC homopolar motors are the simplest in theory but have many thorny problems. They have a great potential for application as ship drives, contingent on resolution of brush reliability/maintainability issues as well as the complexity associated with generation and management of high DC power levels at low voltage. Another potential application could be as wind turbine generators, which have characteristically of low speeds.

REFERENCES

1. A. D. Appleton, "Superconducting DC Machines," *in a book Superconducting Machines and Devices*: Large Systems Applications, edited by S. Foner and B. B. Schwartz, Plenum Press, New York, 1973.
2. M. J. Superczynski and D. J. Waltman, "Homopolar Motor with High Temperature Superconductor Field Windings," *IEEE Trans. Appl. Superconductivity* 7(2): 513–518, 1997. DOI 10.1109/77.614554.
3. M. Heiberger, W. P. Creedon, M. R. Reed, K. M. Schaubel, B. J. O'Hea, and A. Langhorn, "Superconducting DC Homopolar Motor Development for

U.S. Navy Ship Propulsion," Proc. 3rd Naval Symp. Electric Machines, Philadelphia, December 4–7, 2000.

4. R. J. Thome, W. Creedon, M. Reed, E. Bowles, and K. Schaubel, "Homopolar Motor Technology Development," *IEEE Power Engineering Society Summer Meeting*, Chicago, IL, July 25, 2002, 260–284, IEEE ISBN 07-7803-7518-1.

5. L. Petersen, "DOD Perspective," Talk presented at the 2008 US-DOE Peer Review of Electric Delivery and Energy Reliability.

6

SYNCHRONOUS AC HOMOPOLAR MACHINES

6.1 INTRODUCTION

Synchronous homoploar machines have been in use for a very long time, principally for generating power for industrial high-frequency loads from 250 Hz to 200 kHz. These machines have the same terminal characteristics as traditional field-wound synchronous machines. The defining feature of these machines is the homopolar d-axis magnetic field created by a field winding that is fixed to the stator and encircles the rotor rather than being placed on the rotor. The absence of rotor windings allows the speed to be raised to the centrifugal stress limit of the rotor materials. There are several advantages to having the field winding in the stator. Among these are elimination of slip rings and greatly simplified rotor construction, making it practical to construct the rotor from a single piece of high-strength steel. Solid steel rotors make homopolar machines attractive for high-speed operation. A stationary superconducting field winding also becomes attractive because tasks such as mechanical support, cryogenic cooling, and electrical interfaces become much simpler. The superconducting field winding also allows high flux levels to be achieved efficiently, making a slotless stator design feasible. The slotless stator design is similar to that of conventional or permanent magnet machines with the exception of

Applications of High Temperature Superconductors to Electric Power Equipment, by Swarn Singh Kalsi
Copyright © 2011 Institute of Electrical and Electronics Engineers

challenge of bonding the stator coils to the smooth inner bore of the stator iron yoke.

6.2 PRINCIPLE

The simplest configuration of a homopolar synchronous machine is shown [1] in Figure 6.1, along with magnetizing flux path. Such a machine has two stators and two rotors forming a single magnetic circuit, energized with an annular DC excitation winding. It is a sine-wave AC machine and has cylindrical surfaces on both sides of the air gap. The stator is a conventional poly-phase AC stator, while the rotor has internal flux barriers shaped to maximize the ratio of d-axis (high-inductance axis) to q-axis (low-inductance axis) reactance. These machines can be designed to produce torque with low ripple and low acoustic noise due to its cylindrical construction and sine-wave AC operation. By staggering rotor poles by a pole-pitch, it is possible to employ a single stator winding with straight coils. GE has developed such a machine [2] by utilizing superconducting field winding. It is a 1-MW, 15,000-RPM generator and has a solid rotor, a poly-phase stator winding and superconducting field winding. It offers advantages such as high rotor speeds and stationary armature and field excitation coils. The stationary topology of superconducting field coil minimizes risks and simplifies cryogenic cooling system.

Figure 6.1 shows a homopolar machine with a solid rotor with four salient poles. The flux traverses air gap only in radial direction for each stator section (i.e., left stator or right stator). For example, in the left stator, the flux traverses from rotor to stator, and in the right stator, it traverses from stator to rotor. Thus at any radial location the voltages induced in the stator coils are of opposite polarity in the left and right

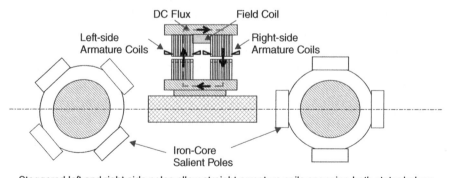

Staggered left and right side poles allow straight armature coils spanning both stator halves

Figure 6.1 Configuration of a homopolar AC machine

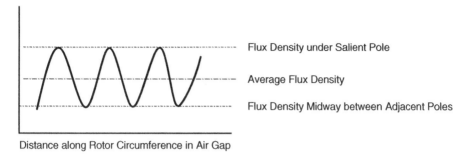

Figure 6.2   Flux density variation in air gap

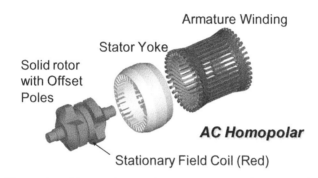

Figure 6.3   Homopolar machine configuration (Courtesy GE)

stators. For this reason it is necessary to split the stator into left and right sections—just as shown in the figure. The flux density in the air gap under a pole (*d*-axis) is much larger than that at midpoint between adjacent poles (*q*-axis). The flux densities under a pole and at midpoint between adjacent poles are shown in Figure 6.2. An average flux density line can be defined about which peak of high and low flux densities is about equal. A four-salient pole flux distribution could be viewed as an eight-pole distribution about the average line. The induced AC voltage in the armature coils corresponds to the eight-pole flux density distribution. The two separate stator coils require an end-turn space between them. This space can be reduced if a single coil spans both stator sections. To achieve this, the left and right set of rotor poles are staggered by a pole-pitch as shown in Figure 6.1. The flux density distribution of Figure 6.2 can now be applied to the stator coil sections occupying active parts of the left and right stators. The excitation field winding is sandwiched between the two stator sections and is housed in the stationary part of the stator as shown in Figure 6.1. An exploded view of a homopolar machine [2] consisting of a stator AC winding, a DC field excitation winding, and a salient pole rotor is shown in Figure 6.3 (courtesy GE).

6.3 DESIGN ANALYSIS

With a known field distribution in the air gap of the machine, the design procedures used for a conventional machine (explained elsewhere [1]) can be employed. Volt-ampere rating of a three-phase machine is given below by

$$S = 1.11 K_w \pi^2 B_g ac D_a^2 L_{act} rps, \tag{6.1}$$

where

K_w = stator winding distribution factor,
B_g = average air-gap flux density (T),
ac = armature current loadings (A/m),
D_a = air-gap diameter (m),
L_{act} = active length of stator (m),
rps = rotational speed of rotor (rev/s).

For the staggered rotor pole configuration, L_{act} is the sum of active parts of the two stator sections. B_g is the field in the air gap as shown in Figure 6.2.

The stator operating voltage, current, and number of phases are determined using approaches used for designing conventional machines, as explained in [1]. The machine's operating parameters could be calculated using the formulations in this reference. However, for more accurate results, 2D (or preferably 3D) finite-element programs should be employed. Once the machine parameters are known, its operational characteristics could be calculated using the conventional analysis approaches of synchronous machines.

6.4 DESIGN CHALLENGES

A homopolar machine consists of a superconducting field excitation coil and a conventional poly-phase armature winding. The armature winding could be housed in iron-tooth laminations as in a conventional machine. This allows the employment of conventional conductors for constructing armature coils and their cooling by air or liquid as necessary. However, the working flux density in the air gap is limited by the saturation of the iron teeth. Since the superconducting field winding can generate a very high field in the air gap, it is possible to eliminate the iron teeth. When the iron teeth are eliminated, it becomes necessary to build stator coils with the fully transposed Litz conductor discussed in Chapter 4.

Field winding consists of a single coil sandwiched between two stator sections. The field coil generates flux lines that enter into the rotor in one-half of the stator and return through the other half as shown in Figure 6.1. The flux lines enclose the field excitation coil as shown in the figure. This requires the addition of iron at the outside diameter of the excitation coil. The thickness of this iron increases as a higher field is generated by the field coil. In addition the space allocated for the field coil has an allowance for accommodating the thermal insulation to keep the field coil at the cryogenic temperature while the surrounding components (stators and iron) are at ambient room temperature. This increases the axial separation between the two stator sections, and in turn increases the leakage flux, namely of the flux that does not return through the rotor and stators. Although the superconducting field coil does not experience any reaction torque, it must bear hoop forces due to self-current and thermal stresses during the cool-down and warm-up cycles. Furthermore it must be mechanically supported at the cryogenic temperature with the support components having one end at the cryogenic temperature and the other end at ambient temperature. These support components must limit the thermal conduction from the warm to cold regions. This coil could employ any superconductor and cooling system discussed in Chapters 2 and 3, respectively.

The maximum rotor speed and its bearing support should be designed using best practices for designing high-speed dynamic systems.

6.5   PROTOTYPES

General Electric has built a 1-MW homopolar generator for the US Air Force [3], as shown in Figure 6.3. Subsequent to this GE conducted a feasibility study [2] for 36-MW, 3600-RPM generator, a 36-MW, 120-RPM propulsion motor, and a 4-MW, 7000-RPM power generation module. Many advantages of homopolar machines resulting from keeping the coil mechanically separate from the rotor and keeping it stationary were noted, as are listed below:

- The stationary coil does not experience the large centrifugal forces. The coil can be a simple solenoid around the rotor instead of a more complicated racetrack type of coil, so the coil support can be much simpler. The thermal insulation between the coil and ambient is improved because of the lack of centrifugal loads and reduced requirements on the coil support.

· Without the large forces and resulting strains in the superconducting coil, more reliable HTS coils can be produced using BSCCO or YBCO coated conductor technology. Further, the reduced ampere-turns required by this machine design enable considerable reduction in the utilization of superconductor compared to air-core machine designs.

· The field coil and its cryostat are stationary. Thus there is no need for a transfer coupling to introduce a cooling medium into the rotating cooling circuit. The coil can be cooled instead by one of the established, more reliable ways of cooling, including conduction cooling. The vacuum or foam insulation, as required for good thermal insulation, will be stationary and therefore highly reliable. Other parts of the insulation scheme can also be made more reliable without the large g forces experienced by field coils on rotor.

· There is no need for a "slip-ring" assembly to transfer current to the coil from a stationary exciter.

· There is no need for rotating brushless exciters.

This study concluded that superconducting homopolar machines offer the highest power density for the high-speed applications.

6.6 SUMMARY

A synchronous homopolar machine offers high power density for operation at speeds of 3600 RPM and higher. The penalty for homopolar flux utilization is compensated by higher air-gap flux densities achieved with superconducting field winding and higher rotor speed. A conventional synchronous machine employs individual superconducting coils for each pole, but a homopolar machine uses only a single field coil with much simpler cryogenic requirements. This enhances reliability of superconducting homopolar machines.

REFERENCES

1. M. G. Say, *The Performance and Design of Alternating Current Machines*, S. Pitman, London, 1958.

2. K. Sivasubramaniam, E. T. Laskaris, M. R. Shah, J. W. Bray, and N. R. Garrigan, "HTS HIA Generator and Motor for Naval Applications," Presented at the 17th International Conference on Electrical Machines, Chania, Greece, 2006.

3. K. Sivasubramaniam, T. Zhang, M. Lokhandwalla, E. T. Laskaris, J. W. Bray, B. Gersler, M. R. Shah, and J. P. Alexander, "Development of a High Speed HTS Generator for Airborne Applications," *IEEE Trans. Appl. Superconductivity*, Vol. 19, Issue 3, Part 2, 2009, pp. 1656–1661, Digital Object Identifier, 10.1109/TASC. 2009.2017758.

7

TRANSFORMERS

7.1 INTRODUCTION

Widespread generation and utilization of electric power would have been impossible without transformers. Transformers are the most widely used equipment in an electric grid. Most of the power is generated as AC at relatively low voltages and is utilized at even lower voltages, but the bulk of power is transmitted from points of generation toward points of consumption at very high voltages. Transformers convert electric power from one voltage level to another. Generally, power is transformed almost 5 to 10 times [1] between points of generation and utilization. A conventional transformer employs normal conductor (copper or aluminum) windings, and a small portion of power is lost in these windings in form of joule heating. Assuming an average efficiency of a transformer to be about 99% (including all sizes), it is possible to lose 5% to 10% of total power in the transformers. This represents a huge wastage of power (or energy), especially during the current era of high energy costs. It is possible to reduce the energy loss in transformers by 60% or more by replacing normal conductor windings with superconductor windings. The prospect of achieving this is looking brighter now because of recent progress in the availability of

Applications of High Temperature Superconductors to Electric Power Equipment, by Swarn Singh Kalsi

high temperature superconductors (HTS), especially the second-generation (2G) YBCO-123 coated conductors that have nearly attained the performance goals necessary for their application in transformers, and their cost appears to be approaching the economic viability.

Some of the significant benefits of HTS transformers are listed below:

- *Smaller and lighter.* Due to the high power density of HTS wire, an HTS transformer could be about 40% to 60% smaller in size and weight than conventional transformers of the same power rating. This feature can help reduce manufacturing, shipping, and installation costs. In addition it provides flexibility in siting by reducing space requirements—a critical factor in applications like overcrowded urban substations, wind turbine nacelle, railroad, and shipboard.

- *Longer life.* Heat generated by electrical resistance and constant changes in temperature (thermal cycling) during operation is one of the factors in the degradation of the electrical insulation on the copper wire used in conventional transformer coils. This effect is eliminated with HTS coils that operate at constant cryogenic temperatures. In addition conventional transformers can be overloaded for only short periods (200% for 30 minutes, according to IEEE/ANSI standards). Thermally independent HTS transformers can carry overloads when needed, with no decrease in equipment lifespan and with manageable additional load losses.

- *Higher efficiency.* Energy losses are reduced by eliminating joule losses in the windings.

- *Environmentally benign.* Inexpensive and environmentally benign liquid nitrogen replaces the conventional oil as the electrical insulation (dielectric) and provides the necessary cooling media. Liquid nitrogen is safe, nonflammable, and environmentally friendly. Using it as a dielectric and coolant instead of oil eliminates a potential source of explosions and the potential for soil contamination from oil leaks. HTS transformers can be designed to meet indoor specifications.

- *System performance.* HTS transformers can be designed to provide low reactances that translate to servicing of load downstream with minimal voltage drop. This feature could reduce or eliminate the need for tap changers, which are bulky, expensive, and unreliable.

- *System security.* HTS transformers offer the potential for improving system security. They can interface directly with underground superconducting cables. With no oil to spill or ignite, HTS transformers present a less explosive target during an attack on a system.

· *Fault current limiting.* Material properties of HTS wire provide fault current limiting capabilities, as discussed next in Chapter 8. It may be possible to include fault current limiting features in transformers for certain applications.

Many transformer development projects [2,3,4,5] have been undertaken and technical success has been demonstrated. The commercial introduction of HTS transformers is a function of the availability of tailored 2G wire with an acceptable performance and price.

7.2   CONFIGURATION

The physical basis of the transformer is mutual induction between two coils linked by a common magnetic flux. For efficient energy transfer between the two coils, a strong mutual magnetic flux is desired. An iron core is usually employed as a conduit to carry the flux between coils. Benefits of an iron core are as follows:

· Increased mutual flux due to lower reluctance of the iron core.
· Lower magnetizing current because fewer ampere-turns are required to establish the needed flux in the core.
· Much reduced leakage flux (the flux that does not link a coil) due to the small cross section of the windings.

These benefits are possible at the expense of iron-core losses. The configuration of a typical conventional transformer is shown in Figure 7.1. A conventional transformer employs copper windings with conductor current densities around 4 to 5 A/mm^2 and active oil cooling with fans. Such low current densities makes the winding's cross section large, which in turn increases the weight of copper. For large transformers the assembly shown in Figure 7.1 is housed in an oil-filled metallic tank. The oil cools the windings and enhances their electrical insulation.

Conventional distribution type of transformers are usually built with wide sheet conductors that are capable of carrying a full rated current. Because of the high room-temperature resistivity of copper, the eddy-current losses due to the leakage field are minimal. The 2G HTS conductor are also manufactured in widths up to 40 mm that can carry significant current. However, because of the nearly zero resistivity of

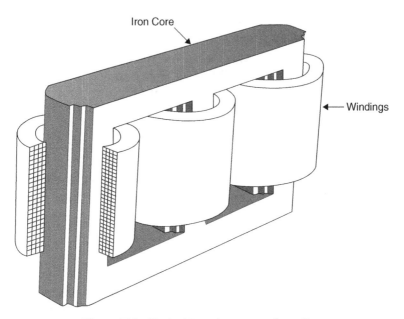

Figure 7.1 Typical transformer configuration

superconductors, large hysteresis and eddy-current losses are caused by the leakage field. During the last 10 years many prototype transformers were attempted with 1G wire (BSCCO-2223), which is typically about 4 mm wide. These attempts were not successful due to the low current-carrying capability of the wire and its high AC losses. Most developers concluded that a low-loss HTS wire is a prerequisite to the development of practical and economically viable transformers. The 2G wire, available from sources in the United States, Europe, and Japan, has intrinsically low loss when oriented parallel to the local magnetic field and is amenable to strategies for reduced perpendicular field losses. The available standard 4-mm-wide wire carries about 100 A and is not suitable for manufacturing large transformers. Wider width (10–40-mm) conductors available as custom orders have a reduced wire current density and generate higher AC losses. They also may not fit standard transformer manufacturing processes. Thus it is necessary to reconfigure the 2G wire for use in transformers. The continuously transposed cable (CTC) technology [6,7] makes this feasible. The CTC cable is like a Roebel conductor commonly employed in large AC generators and large power transformers. Industrial Research Ltd (IRL) of NZ and General Cable Superconductors (GCS) are developing manufacturing processes for CTC using commercially available 2G wire.

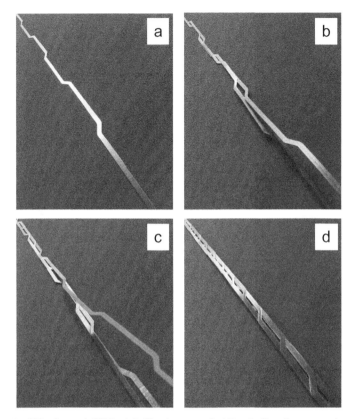

Figure 7.2 Winding of CTC. (*a*) Single strand, (*b*) winding in second strand, (*c*) winding in third strand, and (*d*) completed cable (Courtesy IRL)

Usually a Roebel cable employs a number of wide copper strands bent laterally with a certain periodicity and then assembled such that each strand occupies all possible locations over a given length of the cable. The 2G wire cannot be bent laterally, but it can be cut to form a strand of a Roebel cable. Figure 7.2 shows the steps during construction of a CTC. The cable is constructed using the conventional number of strands/strand width in millimeters. The necessary number of strands can be assembled to achieve a desired current in a CTC. AC losses in a CTC are the sum of losses in individual strands plus a small loss due to magnetic coupling among the strands.

IRL and others are developing formulas for estimation of AC losses by conducting measurements on coils built with CTC [8]. Currently General Cable Superconductors is manufacturing CTC using 2-mm- and 5-mm-wide strands. The transposition pitches for these cables are 90 and 300 mm, respectively. It is possible to assemble 2-mm-wide 10

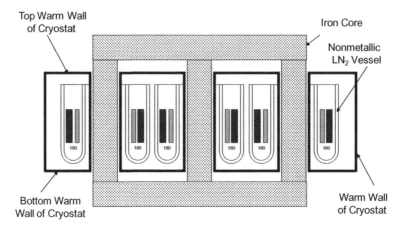

Figure 7.3 Generic HTS transformer configuration

strand (10/2) and 5-mm-wide 15 strand (15/5) CTC conductors. Other combinations are possible with straightforward changes in tooling.

HTS transformer windings are contained in nonmetallic vessels and are operated at cryogenic temperature by cooling with LN_2 or other means. A generic configuration of an HTS transformer is shown in Figure 7.3. The iron core is operated at room temperature because the iron-core losses remain essentially unchanged at the cryogenic temperature and it is not economical to remove them with a refrigerator. The outermost vertical warm wall that encloses the vessels housing the windings and their cryogenic coolant can be metallic. The top and bottom walls (both warm and cold) must be nonmetallic, or they should be segmented to prevent formation of shorted turns around the vertical core limbs. These walls must be designed to withstand the vacuum inside and the atmospheric pressure outside. A single metallic warm wall could enclose windings and iron core to provide protection against the environment as shown in Figure 7.4. A plan view of such a transformer is shown in Figure 7.5.

HTS windings contained in a liquid nitrogen (LN_2) vessel are operated at around 70 K under 3 to 4 bar pressure. A system to achieve subcooled LN_2 in the winding vessel is shown in Figure 7.6. The LN_2 is cooled in a separate cryostat with cryocoolers, using the arrangement shown in the figure, and pumped to the winding vessels. One pump can be used for pumping LN_2 to all vessels. The temperature of LN_2 is controlled by controlling the temperature of cryocooler cold heads.

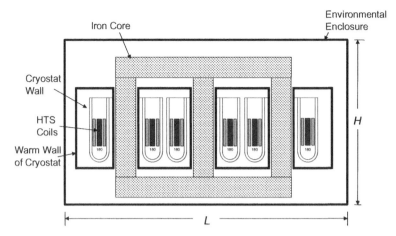

Figure 7.4   HTS transformer configuration with environmental protection

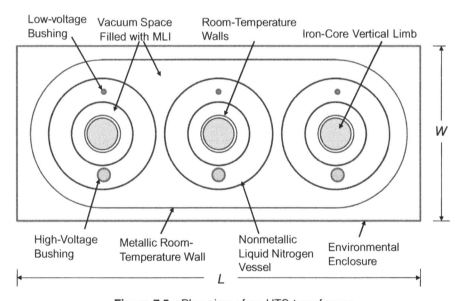

Figure 7.5   Plan view of an HTS transformer

Excessive cooling power will lower the temperature of LN_2, which is preferable from the viewpoint of gaining an extra safety margin to carry the overload. In an actual system the heaters are installed on the cold heads to keep their temperature above the freezing temperature of LN_2 to prevent icing. Such systems have been successfully built and operated by Oak Ridge National Laboratory and others.

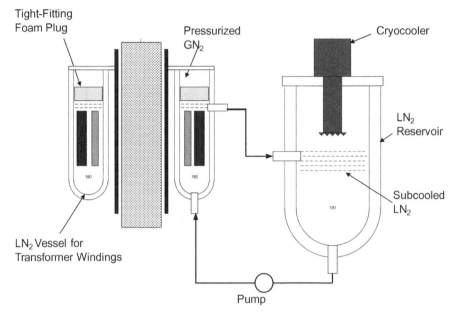

Tight-Fitting
Foam Plug

Pressurized
GN$_2$

Cryocooler

LN$_2$
Reservoir

Subcooled
LN$_2$

LN$_2$ Vessel for
Transformer Windings

Pump

Figure 7.6 Cooling system configuration

7.3 DESIGN ANALYSIS

The basic design process for the HTS transformers is similar to that of conventional transformers [9,10]. A good design is a function of the optimal use of active materials such an iron-core, HTS material and cryogenic cooling system. Baldwin et al. [11] and Berger et al. [12] have published procedures for optimizing design of HTS transformers. A 50-MVA HTS transformer design is presented later in this section for demonstrating the design process. This design may not be optimal in terms of cost, size, weight, or efficiency. A large transformer design is generally optimized on the basis of customer requirements and manufacturer's capabilities. Below are discussed a few key issues that have a significant impact on the size, weight, cost, and performance of an HTS transformer.

- *AC losses in HTS windings.* CTC is used in this design analysis in order to minimize the AC losses in the windings. The AC losses could represent a significant portion of the total thermal load on the refrigeration system, but no reliable analysis is available for estimating these losses while a CTC carries the AC and experiences the external AC field. IRL and others are in process of developing

AC loss analysis formulations. Due to unavailability of good analysis basis, AC losses have not been estimated for the windings.

· *Size and weight.* Voltage per turn is a measure of core limb's cross section. A larger core cross section may lower HTS consumption at the expense of larger weight and size. The larger core will also require bigger diameter coils. However, a manufacturer may prefer winding diameters no larger than what their existing machinery can handle. Thus, by keeping the core diameter similar to that of conventional transformers, it is possible to reduce the overall size and weight of HTS transformers of similar rating. In other words, a transformer manufacturer could produce HTS transformers of twice the rating within the capabilities of their existing winding and handling equipment, and facility space. However, some customers may not mind the larger weight (caused by the larger diameter core), provided that the product price is lower. Selection between the two approaches is better made by discussion between a customer and a manufacturer.

· *Operating temperature.* A conventional oil-cooled transformer designed for 100°C operation can be operated at 50% overload by circulating oil in the tank and at 100% overload by providing additional cooling fans. Similar ratings are also possible with HTS transformers. For example, if an HTS transformer is designed for operation at 77 K, then it is possible to overload it by 50% and 100% by operating it at 70 K and 64 K, respectively. Lower temperature operation will require additional cryogenic cooling capacity.

· *Operational constraints.* Since HTS windings are more compact than copper windings of a conventional transformer, the leakage reactance of HTS transformers can be designed to be low. A low leakage reactance results in lower output voltage variations between no-load and rated-load conditions. It might also be possible to eliminate use of tap changers typically employed to correct output voltage as a function of load. However, lower leakage reactance generates higher through fault currents and forces during a short-circuit event experienced by a transformer. Thus a compromise is needed between lower leakage reactance and acceptable fault current.

· *Compliance with industry standards.* Per IEEE [13] and IEC [14] standards, a category II transformer is expected to withstand a terminal fault for a period of 2 seconds. Once the circuit with fault has been isolated, the transformer is required to keep serving the healthy circuits, meaning the circuits without the fault. In order to satisfy this requirement in an HTS transformer, it may be necessary to co-wind a thick copper (stabilizer) tape with a CTC. However, a thicker copper tape occupies more space and impacts the transformer's size and weight. This can erase any size and weight advantage with respect to a conventional transformer. AC losses in a thicker tape may also be unacceptable. To let a transformer bear load following clearance of a fault, it is essential to limit the temperature of the HTS windings to less than 90 K by the end of a fault. On the other hand, if the windings temperature is higher than 90 K, it would be necessary to allow the windings to cool down to 77 K (normal operating temperature) before letting them bear the load. A possible solution is to use a tape thickness that limits winding temperature to 90 K at end of a 0.25-second long fault. This approach would require that the HTS transformers be protected with circuit breakers or other fast-acting switching devices, including faster means for detecting a fault.

- *Transient currents during switching.* When a transformer is suddenly connected to an electric grid, it can draw a large inrush current for magnetization of the iron core. This current could be more than 10× the normal current and will persist for a long time in an HTS transformer. This high current will definitely quench the HTS winding, which has then to be cooled back to its superconducting state before allowing it to carry any load. This will require significant downtime before the transformer is available for bearing the load. It is possible to minimize the switching current by utilizing a point on the wave-switching scheme. With this approach, the transient current could be reduced to 1% to 2% of the nominal current of HTS windings. This issue has been addressed in the literature [15,16,17].

The basic design equations are presented below.

7.3.1 Core Sizing

The core cross section is selected on the basis of an assumed voltage per turn. The voltage per turn (e_t) is given by

$$e_t = \frac{\omega}{\sqrt{2}} \cdot k_{fe} \cdot \pi \cdot r_c^2 \cdot B_c, \tag{7.1}$$

where

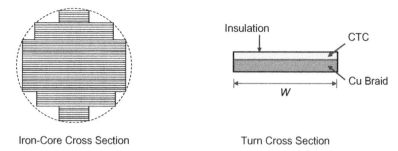

Iron-Core Cross Section   Turn Cross Section

Figure 7.7   Cross sections of iron core and a coil turn

ω = frequency = $2\pi f$,

r_c = apparent radius of core,

k_{fe} = core factor (fraction of iron in the core cross section),

B_c = operating peak flux density in the core.

Once e_t is selected, the core radius (r_c) can be calculated from equation (7.1) for a selected peak operating field in the core. The core factor is available from any transformer design book [9] and is equal to 0.75 for a three-step core shown in Figure 7.7.

Since the HTS transformer must withstand a short-circuit fault for a certain period of time, it is necessary to add copper to individual turns. This can be accomplished by co-winding a copper braid with the CTC turns. A cross section of such a turn is shown in Figure 7.7. The CTC and copper braid are wrapped with turn insulation. Turns in primary and secondary windings are determined by dividing phase voltage by e_t for each winding.

7.3.2   50-MVA Example Design

The conceptual design of a 50-MVA, 132-kV/13.8-kV transformer is presented to demonstrate the design process and to identify component technology and performance issues. Specifications are utilized for a magnetic design of the HTS transformer. The primary winding connected in star configuration operates at 132 kV, and the secondary is delta connected for operation at 13.8 kV. This HTS transformer design is based on the following major assumptions.

1. Both primary and secondary windings employ CTC made of 2G HTS wire. A typical CTC cable is shown in Figure 7.8. The current

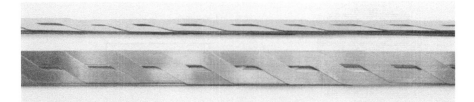

Figure 7.8 Typical CTC made with a 2G HTS conductor (Courtesy IRL)

Table 7.1 Specifications for a 50-MVA HTS transformer

Parameter	50 MVA
Rating, MVA	50
Primary line voltage, kV	132
Primary connection type	Wye*
Secondary line voltage, kV	13.8
Secondary connection type	Delta*
Frequency, Hz	60
Cooling	Liquid N_2
Continuous overload capability	50%
Operating temperature, K	
Normal load	77
Overload	70
Insulation Qualification	
AC voltage with stand, kV	230
BIL (IEEE Std. C57.12.00–2000 Table 4), kV	550

*In the United States it is customery to employ the primary with Delta and the secondary with Wye connections.

    density of each strand (after stamping) is assumed to be 300 A/cm at 77 K.

2. Primary winding employs 2×8 CTC (2 mm wide $\times$ 8 strands) and has a critical current of 480 A_DC in the self-field at 77 K.

3. Secondary winding employs 5×17 CTC (5 mm wide $\times$ 17 strands) and has a critical current of 2550 A_DC in self-field at 77 K.

4. Iron core is similar to that employed in conventional transformers and operates at room-temperature.

5. Primary and secondary windings of each phase are housed in individual liquid-nitrogen vessels.

As shown in Figure 7.4, concentric winding arrangement around each core limb is selected for each phase. In order to minimize the

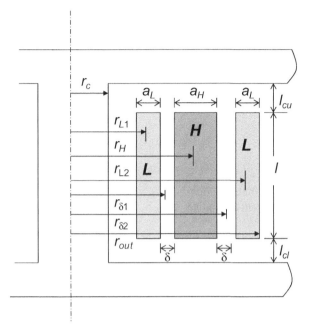

Figure 7.9 Arrangement of primary and secondary windings

leakage flux, the secondary winding is divided into two parts that strad-
dle the single primary winding. The arrangement of windings inside a
core window is shown in Figure 7.9, and the various dimensions labeled
in the figure are listed in Table 7.2 for the 50-MVA transformer example.

A design summary of the 50-MVA HTS transformer is provided in
Tables 7.3 through 7.7. The typical voltage/turn (e_t) for this size transformer is around 90 V-rms/turn. An operating flux density in the transformer is selected to be 1.6 T. On the basis of these two quantities, the
core limb diameter is determined to be 600 mm. Based on these volt/
turn values, the number of turns per phase for the star connected
primary winding is 860 and the number of turns per phase for the delta
connected secondary winding is 156.

Using the symbols and information in Table 7.2, it is possible to calculate distance between adjacent limbs (L_{sp}) of transformer core:

$$L_{sp} = 2(r_{L2} + 0.5a_L) + t_{cso}. \tag{7.2}$$

Window height (L_h) is given using the maximum coil height:

$$L_h = \max(l_H, l_L) + l_{cu} + l_{cl}. \tag{7.3}$$

Table 7.2   Winding dimensions

Description	Symbol	Value (mm)
Core radius	r_c	300
High-Voltage Winding Details		
Radius to midpoint	r_H	402.5
Height	l_H	516
Radial build	a_H	21.5
Low-Voltage Winding Details		
Radius to midpoint of inner winding	r_{L1}	364.5
Radius to midpoint of outer winding	r_{L2}	440.5
Height	l_L	507
Radial build	a_L	8.7
Gap between adjacent windings	δ	23
Clearance between Windings and Core		
Upper	l_{cu}	300
Lower	l_{cl}	300
Radial space between inner LV winding and core	t_{csi}	60
Radial space between outer LV winding and outside of cryostat	t_{cso}	40
Separation between adjacent phase cryostats	t_{ph}	30

L_{sp} and L_h are used for calculating window area and weight of vertical limbs of the iron core.

The primary winding is arranged in 10 layers and each layer has 86 turns. This coil employs a CTC cable with 8 strands, and each strand is 2 mm wide. Each cable is 5 mm wide (bare) and 5.1 mm wide (insulated). A CTC cable example is shown in Figure 7.8. A copper braid of 1.6 mm thickness is co-wound with the CTC. Each CTC cable, 5 mm wide and 1 mm thick, is insulated with 0.05-mm-thick insulation. This insulation could be applied by wrapping several layers of DuPont Nomex or Kapton tapes that have good electrical insulation characteristics while impregnated with LN_2. A layer winding configuration is suggested for this winding. A 0.05-mm-thick solid insulation sheet is suggested for use between adjacent layers. The radial build of the primary winding is 22 mm, and the mean-turn length is 2.53 m. Each primary coil is 516 mm in axial length. The CTC conductor needed for all three primary windings is 6.5 km.

The secondary winding is split in two parts, and each part has two layers. This coil uses a CTC cable with 17 strands, and each strand is

Table 7.3   50-MVA HTS transformer design summary—I

Design Reference	Book Example Value
Nominal Rating	
Rating, MVA	50
Primary line voltages, V	132,000
Primary winding connection type	Star
Secondary line voltage, V	13,800
Secondary winding connection type	Delta
Frequency, Hz	60
Maximum flux density in the core, T	1.6
HTS Conductor Characteristics	2G-YBCO
Critical currents, A/cm	300
PRIMARY WINDING CTC PROPERTIES	
Number of strands	8
Strand width, mm	2
Cable width, mm	5
Cable thickness, mm	0.1
Cable current, A	480
SECONDARY WINDING CTC PROPERTIES	
Number of strands	17
Strand width, mm	5
Cable width, mm	12
Cable thickness, mm	2
Cable currents, A	2,550

5 mm wide. This cable is 12 mm wide and 2 mm thick. A copper braid of 3.4 mm thickness is co-wound with the CTC. This cable is insulated with 0.05-mm-thick insulation applied by wrapping two layers of Nomex or Kapton tape. A 0.05-mm-thick solid insulation sheet is use between adjacent layers. Each half of the secondary winding has a 9-mm radial build and is 507 mm long in the axial direction. This arrangement of turns per layer determines the height of the vertical limbs. The total length of the CTC needed for all secondary windings is 1.2 km.

Both primary and secondary windings are contained in a single liquid nitrogen vessel (also called cryostat). A radial spacing of 23 mm is provided between adjacent coils. This space includes a 3-mm-thick winding mandrel and four to five layers of a 0.025-mm-thick solid Kapton sheet. The Kapton sheet is included to provide BIL voltage withstand capability in conjunction with the LN_2 filled gap. One side of the Kapton film is coated with a 0.0125-mm-thick sticky coating of DuPont's Teflon FEP.

Table 7.4   50-MVA-HTS transformer design summary—II

Winding Dimension	Value
Radius of core, mm	300
Radial thickness of LV winding, mm	9
Radial thickness of HV winding, mm	22
Axial length of primary winding, mm	516
Axial length of secondary winding, mm	507
Gap between coils and core on TOP, mm	300
Gap between coils and core on BOTTOM, mm	300
Gap between adjacent windings, mm	23
Cryostat and LN_2 space between inner LV winding and core, mm	60
Cryostat and LN_2 space between HV winding and RT surface, mm	40
Separation between cryostat of adjacent phases, mm	30
Radius to middle of inner LV winding, mm	364
Radius to middle of outer LV winding, mm	441
Radius to middle of gap between inner LV and HV windings, mm	380
Radius to middle of gap between outer LV and HV windings, mm	425
Radius to middle of HV winding, mm	402
Outside radius of outer HV winding, mm	445
Outside radius of a phase cryostat assembly, mm	485
Voltage/turn, V	88.5

The radial space between the cold secondary winding and the warm core is 60 mm. This space is used for the cryostat cold and warm walls and for the thermal insulation between them. A radial space of 40 mm is allocated between the cold surface of the outer secondary winding and the outside wall of the liquid nitrogen vessel. A 30-mm space is allocated between the outer surfaces of the adjacent phase liquid nitrogen vessels. On the basis of these space allowances and coil radial builds, the center-to-center distance between the adjacent limbs of the core is 1000 mm. The overall core width is 3269 mm, and the total height of the core is 2800 mm. The total weight of the core is 17,071 kg. For example, given a 50% allowance for the weight of HTS windings, liquid nitrogen, and environmental enclosure, the total weight of the transformer is estimated to be 25,606 kg. Total core loss is 12.2 kW.

On a per-unit basis, the combined leakage reactance of primary and secondary windings is 0.067 pu. The magnetizing reactance is 105 pu. If the rated load on the secondary side is served with an 0.8 power-factor (lagging) at the rated voltage, then the primary side voltage and current will be 1.04 pu and 1.006 pu (at 0.76 power-factor [lagging]), respectively.

Table 7.5   50-MVAHTS transformer design summary—III

Design Detail	Value
Primary Winding Details	
Primary winding current, A	219
Cables wound in parallel	1
Number of turns	860
Width of a turn, mm	6
Number of layers	10
Separation between adjacent turns, mm	0.05
Turns/layer	86
Axial length, mm	516
Length of mean-turn, m	2.53
Total length of cable/phase, m	2175
Secondary Winding Details	
Secondary winding current, A	1208
Cables wound in parallel	1
Number of turns	156
Width of a turn, mm	13
Number of Layers	4
Separation between adjacent turns, mm	0.025
Turns/layer	39
Axial length, mm	507
Length of mean-turn of inner LV windings, m	2.29
Length of mean-turn of outer LV windings, m	2.77
Total length of cable/phase, m	394
Reactance Calculation	
Base impedance, Ohm	348
Total leakage reactance, pu	0.067
Magnetizing reactance, pu	105

Although Figure 7.3 shows an optimum arrangement for achieving lowest weight and size, it might be advantageous to employ an individual cryostat for each phase in the beginning. Such an arrangement is shown in Figure 7.10. In this arrangement, cooling for each phase assembly is independent of other phases. This allows each phase to be tested separately before installing it in the final assembly. Individual phases could be tested using an arrangement [18] similar to that shown in Figure 7.11. An arrangement of high voltage bushing and HTS coils is shown in Figure 7.12. Waukesha [18] discovered that off-the-shelf

Table 7.6   50-MVA HTS transformer design summary—IV

Iron-Core Detail	Value
Flux/limb, Wb	0.34
Radius of iron core	300
Cross section of vertical limb, m^2	0.21
Height of vertical limb, mm	1,116
C–C spacing of vertical limbs, mm	1,000
Width of windows, mm	400
Height of yoke (cross-leg),mm	542
Width of yoke, mm	600
Overall width of core, mm	2,600
Overall height of core, mm	2,200
Weight of core, kg	17,071
Total weight of transformer, kg	25,606
Overall length of transformer, mm	3,269
Overall width of transformer, mm	1,570
Overall height of transformer, mm	2,800
Core losses	
Limb losses, kW	4
Yoke losses, kW	8
Total core losses, kW	12
Current lead losses, W	238
Cryostat losses, W	102

Table 7.7   50-MVA HTS transformer design summary—V

Transformer Performance	Value
Secondary side VA rating	1
Secondary load voltage, pu	1
Secondary load current pu	1
Secondary load current PF	0.8
Magnetizing current, pu	0.010
Primary winding voltage, pu	1.042
Primary winding current, pu	1.006
Primary side VA rating	1.042
Primary side PF	0.763
Primary winding amp-turns	189,161
Secondary winding amp-turns	188,406

epoxy resin/paper bushing works well in the HTS transformer where the lower end of bushing is at a very low temperature. These bushings can withstand the LN$_2$ temperature at the lower end without cracking or loss in performance.

Instead of the separate cooling arrangement of Figure 7.6, it may be advantageous to employ individual cooling for each phase using a

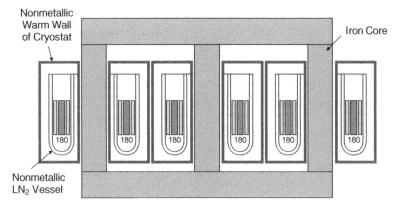

Figure 7.10   Transformer with individual phase cryostats

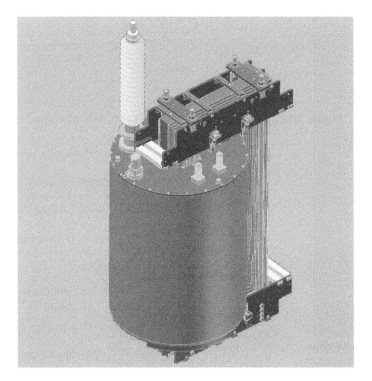

Figure 7.11   A configuration for testing individual phases (Courtesy of Waukesha Electric Systems)

submerged cooling cage inside the cryostat. However, care should be exercised that the cage does not form a closed metallic loop around the core limb to prevent a shorted turn from forming. A possible cooling cage configuration is shown in Figure 7.13. The cage is cooled with a cryocooler that is thermally interfaced with the cage with conduction straps.

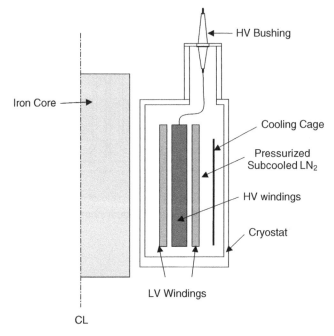

Figure 7.12 Integration of HV bushing with HTS windings

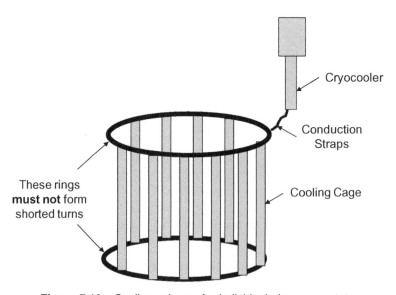

Figure 7.13 Cooling scheme for individual phase cryostats

7.4 CHALLENGES

Transformers of various ratings (few watts to hundreds of MW) are employed in an electric grid. However, due to the space needed for cryostats (thermal enclosures to keep components at cryogenic temperatures), and the cost of HTS wire and refrigeration equipment, it is not economical to build HTS transformers in a smaller rating. The lowest economical sizes could to be around 25 MVA. The lowest possible size is a strong function of costs of cryogenic refrigerator and HTS wire. The HTS wire cost has been projected to be US$25/kA-m in the near future by various manufacturers. Cryogenic refrigerators currently employ G-M cryocoolers (discsssed in Chapter 3), each of which costs about US$30,000. Usually at least two coolers are needed for a transformer. Thus coolers alone add US$60,000 to the cost of an HTS transformer. For example, a conventional 1-MVA transformer priced at US$30,000 may not be an economically attractive size for commercial exploitation.

Most transformer manufacturers prefer to wind coils with a single length of wire; in other words, no joints are allowed in a coil. Thus the availbility of long wire lengths with uniform properties is critical for manufacuring HTS transformers in a factory environment. It is also critical that initial attempts for prototyping HTS transformers be sucessful, lest the technology is viewed by users to be not ready for wide adaptation.

Dielectric properties of electric insulation at cryogenic temperature are also important as partial discharge (PD) activity can weaken the insulation over time. Repetitive cool-down/warm-up cycles can create hidden micro cracks where PD activity initiates and leads to degradation of the insulation system. It is necessary to build experience with an insulation system at cryogenic temperatures.

Like conventional transformers, HTS transformer windings are subjected to high fault forces that are borne by coil cross sections much smaller than those in conventional transformers. Likewise normal and fault forces in the iron core are similar to those in a conventional transformer. It is therefore essential to address these issues in the early stages of transformer design. Key challenges for manufacturing a 50-MVA transformer are the construction of windings with the CTC, LN_2 vessels, cooling system, fault force management, assembly, and operation. To minimize the risk presented by these challenges, it is suggested that demo transformers be built and operated on the distribution grid of a utility. The demonstration transformer should be subjected to all factory and customer approved testing per

IEEE and IEC standards and other applicable standards of the hosting utility.

7.5 MANUFACTURING ISSUES

The technologies needed for manufacturing HTS transformers have already been demonstrated or are in the process of being demonstrated. Below is the status of key technologies needed for manufacturing HTS transformers:

- *HTS wire.* The 2G wire needed for transformers is available commercially from sources in the United States, Europe, and Japan. Its current price is high but is expected to drop to about US$25/kA-m as demand for the wire increases and wire technology improvements are made.
- *CTC conductor.* CTC is available commercially from General Cable Superconductors. The piece lengths are limited by the piece lengths of wire available, or up to 400 m if splices are acceptable. The price and efficiency of CTC manufacture is likely to improve as demand for HTS transformers and other power equipment increases. Besides transformers, CTC can be used for motors and generators, fault current limiters, and a variety of magnet systems.
- *Winding construction.* HTS coils have been built and successfully tested in a variety of devices consisting of motors and generators, transformers, fault current limiters, and magnet systems. Winding fabrication technology has been demonstrated in small to medium-size devices, but it needs work on its adaption for manufacturing in a factory setting. The HTS tapes must be reinforced by adding extra metal to make them robust for withstanding high tension during the winding manufacturing process.
- *Cryostat.* HTS windings are housed in nonmetallic vessels containing liquid nitrogen for cooling. Such nonmetallic vessels have been successfully built worldwide for many transformers and other applications.
- *Cooling system.* Liquid nitrogen cooling systems have been demonstrated in many projects. A closed loop liquid nitrogen refrigeration system can be built using available off-the-shelf cryocoolers. These are modular devices that can be serviced while the transformer remains in operation and keeps serving the load.

- *Iron core.* The iron core to be used in the HTS transformer is the same as that for the conventional transformers. For the same size unit, the HTS transformer will use about half the iron core than that used in conventional transformers.
- *Bushings.* The standard paper/epoxy bushings used in conventional transformers can also be used for the HTS transformers.

Prior to developing capability to build larger units (>50 MVA), it is necessary to first demonstrate the technology in smaller units (1–5 MVA). These smaller units could employ technologies typical of the larger units. Considerable market potential also exists for smaller unit applications in inner city substations, wind generator nacelles, ship systems, and high-rise buildings.

7.6 PROTOTYPES

Waukesha built a full-size HTS transformer under a DOE contract [19]. However, because of a premature dielectric failure, all goals were not realized. Subsequent to this, Waukesha initiated a component development plan, and a full-scale prototype is planned in the future. No other significant prototype development programs are underway at this time.

7.7 SUMMARY

HTS transformer technology is ready for commercial exploitation at this time. HTS wire, CTC, coil construction, and cryogenic cooling technologies have sufficiently matured for practical transformers to be built for utility use. The HTS transformer's economic viability is contingent on achieving a low price wire and refrigeration system at half their current prices. A number of HTS demonstration projects are currently underway globally.

REFERENCES

1. Transformer—Project Fact Sheet, US Department of Energy, Office of Electric Transmission and Distribution, Superconductivity Programs for Electric Systems, Revised May 28, 2004.
2. T. J. Hammons, B. Kennedy, R. Lorand, S. Thigpen, B. W. McConnell, S. Rouse, T. A. Prevost, C. Pruess, S. J. Dale, V. R. Ramanan, and T. L. Baldwin,

"Future Trends in Energy-Efficient Transformers," *IEEE Power Eng. Rev.* Vol. 18, Issue 7, 1998, pp. 5–16.

3. K. Funaki, M. Iwakuma, K. Kajikawa, M. Hara, J. Suehiro, T. Ito, Y. Takata, T. Bohno, S. Nose, M. Konno, Y. Yagi, H. Maruyama, T. Ogata, S. Yoshida, K. Ohashi, H. Kimura, and K. Tsutsumi, "Development of a 22 kV/6.9 kV Single-Phase Model for a 3 MVA HTS Power Transformer," *IEEE Trans. Appl. Superconductivity* 11(1): 2001, pp. 1578–1581.

4. B. W. McConnell, "Transformers—A Successful Application of High Temperature Superconductors," *IEEE Trans. Appl. Superconductivity* 10(1): 2000, pp. 716–720.

5. R. Schlosser, H. Schmidt, M. Leghissa, and M. Meinert, "Development of High-Temperature Superconducting Transformers for Railway Applications," *IEEE Trans. Appl. Superconductivity* 13(2): 2003, pp. 2325–2330.

6. R. A. Badcock, N. J. Long, M. Mulholland, S. Hellmann, A. Wright, and K. A. Hamilton, "Progress in the Manufacture of Long Length YBCO Roebel Cables," *IEEE Trans. Appl. Superconductivity* 19(3): 2009, pp. 3244–3247.

7. N. J. Long, R. Badcock, P. Beck, M. Mulholland, N. Ross, M. Staines, H. Sun, J. Hamilton, and R. G. Buckley, "Narrow Strand YBCO Roebel Cable for Lowered AC Loss," *J. Phys. Conf. Ser.* 97: 2008.

8. L. S. Lakshmi, K. P. Thakur, M. P. Staines, R. A. Badcock, and N. J. Long, "Magnetic AC Loss Characteristics of 2G Roebel Cable," *IEEE Trans. Appl. Superconductivity* 19(3): 2009, pp. 3361–3364.

9. M. G. Say, *The Performance and Design of Alternating Current Machines*, CBS Publishers, New Delhi, 2002.

10. J. H. Harlow, *Electric Power Transformer Engineering*, CRC Press, Boca Raton, 2004.

11. T. L. Baldwin, J. I. Ykema, C. L. Allen, and J. L. Langston, "Design Optimization of High-Temperature Superconducting Power Transformers," *IEEE Trans. Appl. Superconductivity* 13(2): 2003, pp. 2344–2347.

12. A. Berger, M. Noe, and N. Hayakawa, "Design Method for Optimized Superconducting Transformer," presented at the 2008 Applied Superconductivity Conference.

13. IEEE Std. C57.12.00-2000, IEEE Standard General Requirements for Liquid-Immersed Distribution, Power, and Regulating Transformers.

14. IEC 60076-3 and -4, *Power Transformers*.

15. S. J. Asghar, "Elimination of Inrush Current of Transformers and Distribution Lines," Proc. 1996 Int. Conf. Power Electronics, Drives and Energy Systems for Industrial Growth, Vol. 2, January 1996, 976–980, held in New Delhi.

16. US Patent 7095139.

17. T. Ishigohka, K. Uno, and Sakio Nishimiya, "Experimental Study on Effect of Inrush Current of Superconducting Transformer," *IEEE Trans. Appl. Superconductivity* 16(2): 1473–1476, 2006.

18. S. W. Schwenterly and E. F. Pleva, "HTS Transformer Development," Presented by Waukesha at the 2008 US-DOE Peer Review.

19. S. W. Schwenterly and E. F. Pleva, "HTS Transformer Development," Presented by Waukesha at the 2007 US-DOE Peer Review.

8

FAULT CURRENT LIMITERS

8.1 INTRODUCTION

An electric power grid is an extremely complex system consisting of multiple generators, motors, transformers, and switchgear and transmission lines. The electric grid inevitably experiences extreme natural events and faults. For example, an electric short-circuit fault can damage equipment at the location of fault, and the resulting large fault currents can cause highly dynamic and thermal stresses in all grid components. Growth in the generation of electric power and an increased interconnection of the network leads to higher fault current levels. The fault current level could easily grow beyond a system's short-circuit current withstand capability, so it is controlled with a variety of devices. These devices allow equipment to remain in service even if the prospective fault current exceeds the rated peak and short-time withstand current limits. They permit postponement of equipment replacement to a later date. In case of newly planned networks, fault current limiters allow the use of equipment with lower ratings, which render considerable cost savings. The two most commonly used devices to control fault currents are fuses and inductors, though high-impedance transformers also play a major role. Fuses are simple and inexpensive

Applications of High Temperature Superconductors to Electric Power Equipment,
by Swarn Singh Kalsi
Copyright © 2011 Institute of Electrical and Electronics Engineers

fault current limiting devices, but manual intervention is required to replace them following a fault, resulting in a prolonged interruption of power. Moreover fuses are not practical at transmission level voltages because the time required to interrupt a fault is too long. However, passive devices like series inductors connected in series in a line limit the fault current by their impedance. This impedance also limits power transfer over a line during normal operations. It thus impedes quick recovery of a system following a fault and increases the reactance over resistance ratio, X/R, of a transmission or distribution line. The latter disadvantage has consequences on the transient recovery voltage (TRV) ratings for breakers. The higher the X/R ratio will produce a higher TRV on the breaker's opening; hence it will require more sophisticated and expensive breakers. For this reason in-line inductors are employed reluctantly. The best solution is to have a fault current limiting device that reacts very rapidly, resets itself after a fault, and has minimal impact on system performance during normal operation.

A superconducting fault current limiter (S-FCL) appears to satisfy these requirements. The principle of the S-FCL is simple. A superconducting element is inserted in series with a line. During normal operation the system operates without any limitations because the resistance of the superconducting element is essentially zero, and it is possible to minimize the inductive impedance. However, during a fault when the fault current reaches many times the rated value, the superconducting element reverts rapidly to its normal state (i.e., its resistance increases to a defined value). The increased resistance/impedance limits the fault current to the desired level. Once the fault clears, the superconducting coil carries the system's normal current with zero resistance. During a fault when the superconducting element is in normal state, it warms up to a predetermined temperature. The reset time after a fault is a function of time required to cool the superconducting element to its prefault temperature. This is a challenging task.

S-FCLs based on high temperature superconductors (HTS) have been explored since the late 1980s [1,2,3], but a cost-effective, practical, and reliable concept has remained elusive. Ideally an S-FCL should limit the fault current to a desired level during a fault and recover quickly after the fault has cleared. An S-FCL that satisfies the user-defined requirements [4,5,6] has become a holy grail for scientists and engineers working in this field. Several small-scale demonstrations have been conducted [7,8,9], and a few FCL developments have started [10] recently with support from US Department of Energy (DOE). The first commercial S-FCLs have also been shipped recently [11].

8.2   PRINCIPLE AND CONFIGURATION

As explained earlier in Chapter 2, a superconductor has the following two unique properties:

- Zero resistivity to DC current flow below a critical temperature (T_c), critical magnetic field (B_c), and a critical current density (J_c) or critical current (I_c)
- Once the critical surface spanned by the superconductor's material parameters B_c, J_c, and T_c are exceeded the resistivity of the superconducting material increases rapidly

Most S-FCL concepts exploit this sharp transition of HTS from zero resistance at normal currents to a finite resistance at higher currents. The higher resistance of HTS automatically limits the fault currents when they experience currents higher than normal circuit currents. Thus an S-FCL is a self-triggering, fail-safe device as long as the superconductor portion is capable of handling the fault currents and returning to the superconducting state. However, before reaping the benefits of S-FCLs, it is necessary to address the following significant challenges:

- The superconductor component of the FCL (in the resistive type of design) must be cooled to a cryogenic temperature (LN_2 temperature or below), and refrigeration cost must be minimized.
- A normal bypass element (e.g., stainless steel) in parallel with the HTS element is needed to prevent formation of localized hotspots and to act as thermal buffer to absorb local heating.
- AC losses (hysteresis and eddy current) in the HTS material, which increase the cooling cost, need to be managed.
- The high-voltage design must be managed in a cryogenic environment.

So far two material options have proved their capability for reliable limitation of fault currents. The one is melt cast processed $Bi_2Sr_2CaCu_2O_8$ (MCP BSCCO 2212) with a T_c of 92 K, which can be used as casted elements (in form of rods, hollow cylinders, or coils). With this material the first technical viable limiter projects on the medium voltage scale have been realized [8,9]. Another attractive material option is thin films of $YBa_2Cu_3O_{7-\delta}$ (YBCO-123) with the same T_c of typically 92 K. This material preferentially is used as a coating on a tape substrate (commonly known as coated conductors or 2G (second-generation) wires.) Chapter 2 describes the properties and manufacturing processes for both materials. Some manufacturers also use BSCCO-2212 material typically as cast elements.

Actually there are multiple types of FCLs based on HTS. These include the S-FCLs described above, based on the superconductor-to-resistive transition. They also include several types of FCLs using an inductive principle with a coupling to a ferromagnetic yoke. Thus HTS-FCLs could be divided into two categories.

- *Resistive FCL (R-FCL).* Current is limited by increased resistance of HTS in normal state:
  - Noninductive coils
  - Straight wire bundles
  - Bulk rod or hollow cylindrical elements
  - Bulk hollow cylinders cut to mono- or bifilar coils
- *Inductive FCL.* Current is limited by insertion of inductance triggered with an HTS coil:
  - Shielded iron core
  - Saturated iron core

R-FCLs can be further divided in to those types that limit current by virtue of the resistive state of the superconductor after the quenching is initiated by a fault, and those types that employ the HTS component as a fast switch so that the actual current limiting is mostly provided by a parallel reactor or resistor.

Prototypes of some of these concepts have been built and tested with varying degrees of success [7,8]. Some large prototypes for high-voltage application are currently under construction, and some commercial units for distribution voltage have been delivered [11]. A general description of these FCLs follows.

8.2.1 Resistive Fault Current Limiters (R-FCL)

HTS elements are inserted in series with the line being protected. During a fault, I_c (i.e., J_c) is surpassed, and FCLs' resistance increases rapidly, leading to quenching of HTS elements before the first peak of short-circuit current is reached. In 50- or 60-Hz AC systems, the HTS elements quench within 1 to 2 ms after initiation of a fault, depending on the ratio of prospective fault current to normal current. A R-FCL requires a pair of current leads for each phase, which connect the HTS elements at cryogenic temperature to room-temperature bushings. The leads are usually made of normal conducting metal such as bronze or copper, and they generate joule heating as well as conducting thermal energy from warm bushings to cold HTS elements. This adds to the thermal load of the refrigeration system and increases its size and cost. To achieve adequate voltage standoffs, the dimensions of the bushings and cryostat are substantial, particularly at transmission voltages. It is also preferable to isolate the R-FCL after an irreversible quench by a conventional mechanical switch, since the material heats up during the limiting phase and it is usually difficult to recover the superconductivity of HTS elements under load. The need for a mechanical switch compromises the fail-safe nature of R-FCL operation, and multiple switches may be required to achieve redundancy. However, in a few cases, recovery under load has been reported, in the event of which the mechanical switch would not be necessary.

After some recovery time (typically in range of many seconds), during which cooling system cools HTS elements down to their pre-fault operating temperature, the R-FCL is inserted back into the circuit. It is possible to trigger the fast switch with the current in the shunt inductor. The AC field of the shunt inductor is utilized to repulse a plate. This movement in turn is utilized to open the fast switch located in the HTS circuit (see Figure 8.9). This concept has been demonstrated in Japan [12,13]. High-voltage vacuum circuit breakers have also been developed in Japan [14,15]. These breakers could also be triggered by the repulsive force of the shunt inductors.

However, HTS elements designed to limit currents within their flux flow resistivity region could recover to normal operating temperature while carrying rated load currents. Such designs require a large amount of HTS material in order to limit the fault current and enable a quick recovery to the superconducting state following a fault clearance. This increases the AC losses under normal operation and makes them bulky and expensive.

The HTS wire could be utilized for making fault current limiting elements in the form of the wound noninductive (bifilar) coils, as shown schematically in Figure 8.1. In a bifilar coil, the adjacent turns carry current in opposite direction. This arrangement minimizes the stored flux that leads to lower residual inductance and lower standby AC losses of the HTS elements. Alternatively, it is possible to achieve a similar effect by interleaving solenoid coils such that any two adjacent solenoids carry currents in the opposite direction. In addition, pancake coils could be arranged such that the field produced by a pancake coil opposes the field due to its adjacent neighbors. These alternate

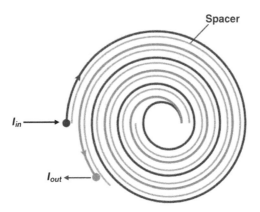

Spacer

I_{in} ———→

I_{out} ←———

Figure 8.1 Bifilar noninductive coil

arrangements store somewhat more flux than the bifilar coil and there-fore have somewhat higher inductance and higher AC losses. Voltage and current requirements are met by connecting the bifilar coils in series/parallel configurations. Individual turns of a coil could employ a single HTS tape or the continuously transposed cable (CTC) described in Chapter 7. The total number of coils could be reduced with the use of CTC because it can carry a much larger current than a single HTS tape. For example, a CTC cable with 15 strands of 5-mm width could carry 1250 A_rms (steady state), which is sufficient for many systems. However during a fault the CTC will be required to carry 5× (= 6250 A-rms) for about 0.1 second. Ideally the R-FCL must recover to its pre-fault state within a few seconds after the fault has been cleared. This is a challenging requirement. Figure 8.2 shows a possible arrangement for coils for all phases housed in a single cryostat and cooled with a common cooling system. It is also possible to house individual phases in a sepa-rate cryostat per needs of a particular application.

Straight R-FCL elements can also be made by stacking necessary number of 2G tapes in parallel, as shown in Figure 8.3. These stacks could employ CTC cable, which works better than a simple 2G tape stack because individual strands of CTC share the current more uniformly. The necessary number of these elements could be arranged in series/parallel configuration to meet voltage and current requirements.

The straight element could also be a simple solid superconductor rod. The necessary number of these bulk rod elements is connected in a series/parallel configuration for a given application. Sometime hollow tubes with a double helix cut into their walls [8] are employed. The current can flow through the walls to generate the bifilar coil effect.

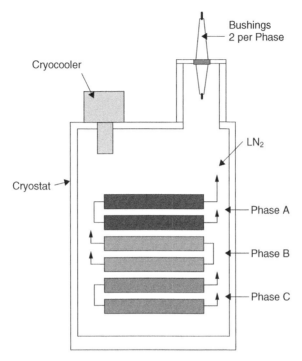

Figure 8.2 FCL assembly with bifilar noninductive coils

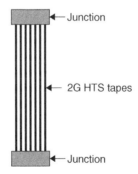

Figure 8.3 FCL element made from 2G tapes or CTC

Such a rod element is shown in Figure 8.4. All straight rod elements could be housed in a single cryostat to create a R-FCL assembly, as shown in Figure 8.5. A practical implementation with bulk rod elements is shown in Figure 8.6.

A prototype with bifilar 2G-coils is under construction in the United States and several with bulk coils have been prototyped in Europe [11].

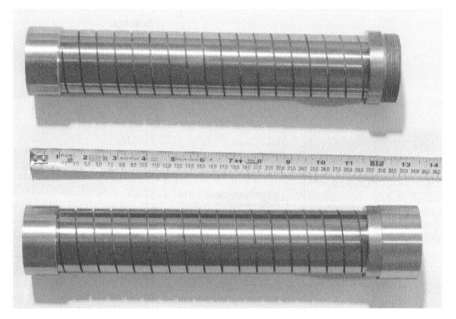

Figure 8.4 Monofilar coils cutout of bulk BSCCO2212 tubes soldered to a metallic shunt (Courtesy of Nexans Superconductors)

8.2.2 Inductive FCL with Shielded Iron Core

A shielded iron-core FCL consists of an iron core, a primary (normal conducting) winding, and a secondary winding made of a superconductor cylinder. This device is like a transformer with shorted secondary windings made of bulk superconductor. During normal operation, the primary winding creates flux in the iron core. But this flux is opposed by the current induced in the secondary winding made of bulk superconductor cylinder. This shields the iron core, and the net flux in the iron core theoretically remains zero. In this state the effective impedance of the primary winding is very low. In fact the impedance is equivalent to the net impedance of a transformer with secondary coil short-circuited. This impedance is equal to the combined leakage impedance of primary and secondary windings and is about 5% to 10% (or 0.05–0.1 pu). During a fault the shielding effect of superconductor cylinder is lost because the large fault current quenches the superconductor. The net current in the superconductor elements in the quenched state is only a small fraction of the fault current in the primary winding. The flux is set up in the iron core once the shielding effect of superconductor has been lost. A large flux linking the primary coil increases its impedance to a predesigned high value, and the fault current is

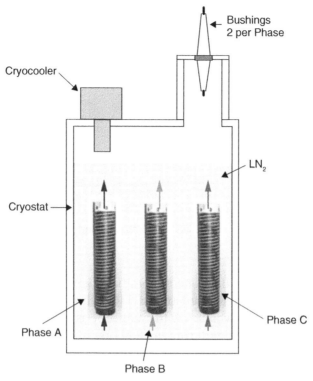

Figure 8.5   FCL assembly employing straight elements

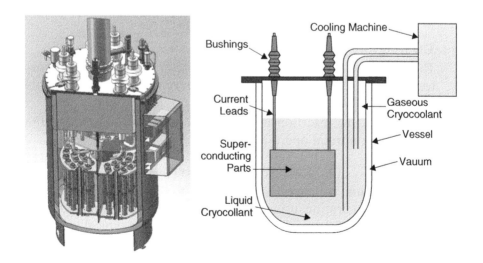

Figure 8.6 FCL assembly employing bulk rod elements (Courtesy Nexans Superconductors)

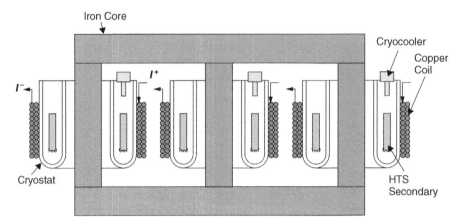

Figure 8.7 Inductive FCL with shielded iron core

limited to the desired value. The main advantage of this system is that the secondary coil is a shorted turn (HTS cylinder) and requires no current leads. The drawback of the concept is insertion of a finite impedance in the line even during normal operation, plus the large size and weight of the transformer core. The device design is in fact similar to a transformer design where the primary winding must support a significant fraction of rated voltage without saturating the iron core. Thus it tends to be large, like the transformer shown schematically in Figure 8.7. After some initial technically successful demonstrations, this concept was abandoned for the following three reasons:

· The iron core makes the system bulky.
· The challenge of making the superconductor cylinder quench uniformly and prevent its failure due to uneven heating and nonuniform thermal stresses.
· Acceptably low insertion impedance is achieved only with difficulty during normal operation.

Although it is possible to replace the solid HTS cylindrical secondary with a CTC* wound coil, but it is still difficult to achieve low impedance during normal operation using the conventional transformer design techniques. The impedance during normal operation is typically similar to that of a transformer (i.e., 0.05–0.1 pu), which most users are reluctant to accept.

*Described in Chapter 7 on transformers.

8.2.3 Inductive FCL with Saturated Iron Core

The saturated iron-core concept utilizes two iron core per phase as shown in Figure 8.8. A conventional copper coil could be used to saturate the cores during normal operation. However, in order to reduce I^2R losses in the copper coil and to make the device acceptable to the users, developers have opted to use a superconducting coil for saturating the core. The most attractive feature of this FCL is simplicity and a fail-safe mode of operation. Faults of long durations can be handled and recovery from a fault is instantaneous, enabling the device to handle multiple successive faults in rapid succession, such as auto-recloses on a protected line or circuit breakers with existing reclosing logic. Explained below is the principle of operation:

- During normal operation, large ampere-turns created with DC in the secondary superconducting HTS coil drive the core into saturation. This lowers impedance of the copper coil in the primary AC side near to that of an air-core coil.

- During a fault, a large fault current demagnetizes the core and drives it from the saturated to unsaturated state (linear B-H region). This increases the primary AC coil impedance. The increased impedance limits the fault current to the desired level.

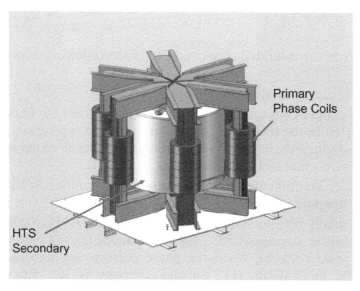

Figure 8.8 Inductive FCL concept with saturated iron core (Courtesy Zenergy Power)

Since an AC wave has both positive and negative peaks to magnetize the iron core, it becomes necessary to employ two separate cores for each phase. Each core has a normal (copper) coil in series with the line being protected. One core works with the positive peak of the AC and the other with the negative peak. A three-phase arrangement of this concept is shown in Figure 8.8, which has six primary copper coils (two for each phase) and a common secondary DC HTS coil for saturating all cores simultaneously. This device, installed in the Avanti Circuit of Southern California Edison in March 2009, became the first SFCL to operate in a US utility system. Abbott [16] has described operation of such a limiter.

Listed below are many desirable attributes of this FCL design:

- Passive and immediate triggering.
- Passive and immediate full recovery after a fault.
- Fail-safe operation.
- Superconductive state maintained at all times.
- Superconductor coil and its associated cryogenics remaining near ground voltage.
- Maintained grid selectivity
- High voltage design similar to that of standard transformers. Each copper coil nevertheless experiences significant system voltage during a fault (determined by the voltage divider effect described in Section 8.3.)

However, the saturated core concept has the following drawbacks:

- Significant normal insertion impedance.
- Two copper coils required per phase. This increase the impedance inserted in the protected circuit during normal operation. Like a transformer, these coils must withstand significant system voltage during normal and fault operations.
- Bulkier assembly due to the six iron cores, resulting in large size and weight.
- Essentially zero total flux experienced by the DC HTS coil when currents in all primary AC coils are balanced (i.e., all phases carry the same current). When the phase currents are unbalanced, a net AC flux links the HTS coil and generates AC in it. This creates AC voltage across coil terminals and induces losses in the HTS coil. The induced AC ripple current and losses are small during

normal operation but are significant during a fault (see Section 8.3.2.)

An HTS coil built with the CTC cable will experience lower induced AC voltage (due to smaller number of turns) and lower AC losses due to smaller strands. Still the main disadvantage of this concept is its large size and weight.

8.3 DESIGN ANALYSIS

Typically, in an electric grid application, an FCL is required to reduce the unlimited fault current to a lower limited value that can be handled by existing circuit breakers and other ancillary equipment. The unlimited fault current is equal to the bus voltage divided by the grid impedance (X_g). To reduce the fault current to a desired limited value, additional impedance (X_{fcl}) must be inserted by a FCL in the circuit being protected. The impedance added by a FCL is defined by the equation below.

$$X_{fcl} = \frac{V_{ph}}{I_{lim}} - X_g,$$

where V_{ph} in the phase voltage of the bus at the point of insertion of the FCL. During a fault the entire phase voltage is experienced by the series combination of X_g and X_{fcl}, which makes the voltage experienced across the FCL circuit considerably less than the rated phase voltage of the bus. For example, to reduce an unlimited fault current of 8 pu to a limited current of 6 pu, the FCL must introduce X_{fcl} equal to $X_g/3$. In this case the voltage drop across the FCL impedance (X_{fcl}) is only 0.25 pu, where 1 pu voltage equals the phase voltage (V_{ph}). The reduced voltage experienced by the FCL leads to a lower HTS material requirement and a smaller size FCL assembly.

   The basic principles for designing all types of R-FCLs are relatively straightforward. A R-FCL limits the fault current by its increased resistance when the HTS wire transitions to its normal state during a fault. The key parameters of a R-FCL design are fault current (I_{lim}), fault duration (Δt) and permissible temperature rise (ΔT) of the HTS elements. The following equations relate these variables. To simplify the analysis, the heat capacity and resistivity are assumed to be temperature independent. The limiting resistance (R) and limiting current (I_{lim}) are given as

$$R = \frac{V_0}{I_{lim}} = \frac{\rho L}{tw}, \quad I_{lim} = tw\sqrt{\frac{C_p \Delta T}{\rho \Delta t}}, \qquad (8.1a)$$

and the corresponding maximum electric field allowed during limitation is independent of the cross section:

$$E_{lim} = \sqrt{\frac{\rho C_p \Delta T}{\Delta t}}, \qquad (8.1b)$$

where

R	= FCL resistance during fault,
V_0	= system rms voltage,
L	= length of HTS current limiting elements,
ΔT	= maximum permissible temperature rise,
Δt	= maximum fault duration (hold time),
ρ, t, w	= resistivity, thickness, and width of HTS,
C_p	= effective specific heat of HTS and stabilizer,
I_{lim}	= maximum current allowed through HTS elements,
E_{lim}	= maximum electric field during limitation.

The minimum conductor volume is obtained by solving these equations. As pointed out by Tixador [3], in the adiabatic approximation the required minimum conductor volume (Vol) is independent of conductor resistivity.

$$\text{Vol} = \frac{I_{lim} V_0 \Delta t}{C_p \Delta T}. \qquad (8.2)$$

For example, a delta connected, R-FCL sized for 15-kV-line voltage, 1250-A/phase nominal current, 4000-A/phase limited fault current for 0.1 second and a 100 K maximum temperature rise during the fault, and with the approximate volume specific heat of 2×10^6 J/m³K, requires 30,000 cm³ of material per phase in all cases. In coil environments it takes many minutes for the coil to cool back to 77 K after a 100 K temperature rise, which sets a time limit for recovery with this approach. Implications of these constraints are discussed below for R-FCLs employing bulk rod and wound coil approaches.

Bulk HTS Rods The R-FCL elements made of bulk rods quickly attain resistivity of $\geq 100\,\mu\Omega$cm during a fault if the current surges far above the critical current of the material. The minimum HTS cross section tw $(= 0.44\,\text{cm}^2)$ is calculated with equation (8.1a) (using $J_c = 4\,\text{kA-dc/cm}^2$ at an operation temperature of 65 K) to carry steady-state current of 1250 A-rms. To prevent hot spots, Nexans employs a metallic shunt (of cross section equal to the HTS cross section). During the limiting phase, a voltage gradient of 0.5 V/cm is achieved, calculated with equation (8.1b). These elements limit the fault current to the desired limit of 4000 A-rms. The total material requirement is 30,000 cm^3 but only half of that is the bulk HTS. This design procedure has led to the design of technically viable prototypes FCLs employing bulk BSCCO 2212.

Coil Wound with the 2G Coated Conductor 2G (YBCO coated conductor) wire could be laminated to a thick (e.g., 100 micron) stainless steel or another high-resistivity, high-strength stabilizer. Such wire architecture is mechanically robust and electrically and thermally stabilized. During a fault, the fault current shunts to the stabilizer and heats it adiabatically. From the previous equations, the minimum conductor volume needed is 30,000 cm^3, and with this stabilized configuration, most of this material can be stainless steel stabilizer. With the 0.5×4-mm^2 stabilizer and 50-μm-thick substrate, 68 km of 2G wire are required. The projected 2G wire cost in the range of $5 to $10/m would be $340 K to $680 K. However, as explained earlier in this section, with the requirement to reduce the fault current from its current values to a slightly lower value, the withstand voltage during a fault is dramatically decreased. This reduces the amount of HTS wire required.

Until recently, the lack of availability of 2G HTS wire was the main impediment to realizing a practical R-FCL. However, long lengths of 2G wire are available now. A technology to laminate the needed stabilizer to 2G FCL wire has been developed, and attractive mechanical properties [17] have been demonstrated.

Linear Elements Made with the 2G Coated Conductor Linear elements [18] could be employed to construct a R-FCL by paralleling the necessary number of tapes to carry the required current. An inductor made of normal material (also cooled by LN$_2$ bath) is connected in parallel with each element. Usually once a R-FCL performs a limiting action, the HTS elements warm to a temperature much higher that their critical temperature. The R-FCL is usually isolated from the load until it has cooled down to its nominal operating temperature. However, by judicious choice of inductance, 2G tape properties, and LN_2 bath temperature and pressure, the developer aims to achieve recovery under load. Nevertheless, the total 2G wire requirement is still similar to the coil version, meaning $30,000\,cm^3$.

8.3.1 Example Design—Resistive FCL

Designing a R-FCL using bulk materials is not simple as it involves dimensioning of the HTS elements to suit the needs of a given application. However, in designing a wire-type R-FCL, it is possible to separate the design functions of the wire and FCL. In principle, a standard 2G wire available from manufacturers can be used in as-is condition or laminated to a high-resistivity substrate. The design example described below is based on a coil type of limiter. The linear element version is similar, with a difference that the 2G wire is assembled into linear element (several meter long) instead of winding into a coil.

The standard 2G wires available from manufacturers are 4mm or 12mm wide, and their current-carrying capability is limited. To carry the large current for a real application, it is necessary to parallel many wires, which makes the FCL design complex and bulky. The example design discussed uses continuously transposed cable (CTC) (described in Chapter 7) built from 2G wire available from AMSC and SuperPower. Table 8.1 lists assumed properties of these wires.

R-FCL devices are sized to the specifications listed in Table 8.2. Each FCL is sized for a 13.8-kV distribution grid and carries 3kA-rms during normal operation. During a fault (short-circuit) the grid circuit experiences a fault current of 40kA. It is desired that this fault current be

Table 8.1   HTS 2G coated conductor wire properties used in the example design

Parameter	AMSC	SuperPower
Overall thickness of YBCO tape, μm		
YBCO layer, μm	1	1
Nickel tungsten, μm	75	None
Hastelloy, μm	None	100
Silver, μm	2.5	2
Critical current density, A/cm	300	300
Additional material applied for FCL		
Stainless steel, μm	120	127
Solder, μm	20	None
Overall strand thickness, μm	218	230

Table 8.2   Specifications for a 13.8-kV, 3-kA fault current limiter

Parameter	Value
Rating, MVA	72
Line voltage, kV	13.8
Line current, kA	3
Unlimited fault current, kA	40
Limited fault current, kA	30
Frequency, Hz	60
Fault hold-time, s	0.1
Conductor type	*CTC*
Number of strands	17
Strand width, mm	5

limited to 30 kA for a period of 0.1 second. The FCL concept shown in Figure 8.9 utilizes an HTS FCL device in parallel with a conventional room-temperature inductor.[†] Although most fault durations are less than 0.1 second, longer duration faults have been experienced, and the system must be capable of handling very long low-level faults, called "through-faults," without switching. To address faults longer than 0.1 second, a fast circuit breaker[‡] is included in series with the HTS FCL element to isolate it at the end of 0.1 second and allow it to cool down to its pre-fault normal operating temperature before inserting it back into the circuit. A CTC consisting of 5-mm-wide, 17 strands (5×17) is assumed in this example design. Other strand configurations might be available in future.

The key analysis assumptions are summarized below:

1. During a fault of very short duration, it is not possible to remove significant thermal energy from the conductor's surface. Because of this limitation, the analysis is based on the adiabatic temperature rise of CTC.

2. In practice, a FCL must be able to carry some overload current before triggering. An overload factor of F_{oc} ($= 1.6$) is assumed.

3. HTS elements of the FCL are designed to fully quench at $2 \times I_c$. Accordingly, a quench factor (F_q) of 2 is assumed. Thus the fault

---

[†] SuperPower places shunt inductor inside LN_2 bath and claim that their design can recover under the load based on the prototype elements tested to AEP reclosure sequence.

[‡] The fast switch can also be triggered using the AC field of the shunt inductor during a fault, as explained earlier in Section 8.2.1.

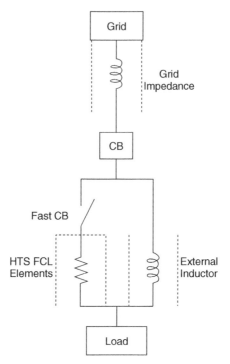

Figure 8.9   HTS FCL concept with a shunt inductor

current in the HTS elements is limited to $I_{lim} = F_{oc} \times F_q \times I_o$, where I_o is the nominal current of circuit being protected.

4. Maximum temperature of the CTC conductor is limited to $300\,K$ in order to avoid damaging FCL coils by thermal stresses. To achieve this, it is necessary to include additional material in the strands of CTC.

The design process described below is based on the SuperPower wire. A similar procedure could be used for designing with American Superconductor Corporation wire.

The first step is to calculate circuit parameters in Figure 8.9. The grid reactance (X_g) is calculated using the phase voltage (V_{ph}) and unlimited fault current (I_{unlim}) specified in Table 8.2:

$$X_g = \frac{V_{ph}}{I_{unlim}}. \tag{8.3}$$

The calculated value of X_g is 0.199 ohm.

Once the HTS FCL branch has been isolated after limiting the fault current for the specified hold period, the total current in the circuit is limited to the through-fault value ($I_{tf} = 30\,\text{kA}$) specified in the Table 8.2. Thus just before the HTS circuit is isolated, the total circuit current is I_{tf}. To satisfy this condition, the circuit in Figure 8.9 has two unknown variables: resistance of the HTS FCL element (R_{fcl}) and reactance of shunt inductor (X_{sh}). Values for these variables are obtained by solving the following two equations (8.4). Calculated values of R_{fcl} and X_{sh} are $0.216\,\Omega$ and $0.073\,\Omega$, respectively.

$$\left| \frac{V_{ph}}{jX_g + \dfrac{R_{fcl}jX_{sh}}{R_{fcl} + jX_{sh}}} \right| = I_{tf}, \quad R_{fcl}I_{lim} = \sqrt{I_{tf}^2 - I_{lim}^2}\,X_{sh}. \tag{8.4}$$

The HTS FCL coil is designed to provide a resistance of R_{fcl} at end of fault hold time ($\Delta T_{hold} = 0.1\,\text{s}$) and at maximum allowable temperature ($T_{max} = 300\,\text{K}$).

The number of strands (N_{par}) needed in parallel to carry the rated current of the circuit being protected is given by

$$N_{par} = round\left(\frac{I_{sco}\sqrt{2}}{I(\theta_{op})/F_{oc}} \right), \tag{8.5}$$

where

I_{sco} = nominal circuit current (A-rms),
$I(\theta_{op}) = I_c$ of an HTS strand (= 197 A for a 5-mm-wide strand),
θ_{op} = nominal operating temperature (= 72 K here),
F_{oc} = overload factor (= 1.6 here).

The total number of strands calculated from the equation above is 34, which translates into two CTC cables. Each cable has 17 strands of 5-mm width.

Each HTS strand consists of many different elements listed in Table 8.1. In addition to these, a stainless steel stabilizer of 127 μm thickness must be used in parallel with each strand. The total effective resistance (ohm/m) of a stabilized strand is given by

$$R_{eff}(I, \theta) = \frac{1}{[1/R_m(\theta)] + [1/R_{HTS}(I, \theta)]} \tag{8.6}$$

where

R_{eff} = total effective resistance of stabilized strand,

R_{HTS} = resistance of HTS layer in a strand,

R_m = resistance of nonsuperconducting elements,

I = current in the strand,

θ = strand temperature.

Once temperature exceeds T_c (critical temperature of HTS), the resistance of the HTS element (R_{HTS}) increases to a very high value. It can be estimated with

$$R_{HTS}(I,\theta) = 10^{-6}\frac{V}{cm} = \left(\frac{I}{IL(\theta)}\right)^N \frac{1}{I}, \tag{8.7}$$

$$IL(\theta) = 2I_{csf} - \frac{2I_{csf}}{20}\left(\frac{\theta}{K} - 70\right), \tag{8.8}$$

where

N = the exponent of the V-I curve for HTS wire,

$IL(\theta)$ = critical current of HTS wire at temperature θ (<90 K)

I_{csf} = critical current is the self-field at 77 K (A/cm)

For all temperatures above 90 K, $IL(\theta)$ is set equal to zero.

The resistance R_m (ohm/m) of nonsuperconductor elements is given by

$$R_m(\theta) = \frac{1}{[1/R_{ss}(\theta)] + [1/R_{ag}(\theta)] + [1/R_{sol}(\theta)] + [1/R_{hst}(\theta)]}, \tag{8.9}$$

where

$R_m(\theta)$ = resistance of all nonsuperconductor elements at temperature θ,

$R_{ss}(\theta)$ = resistance of stainless steel stabilizer,

$R_{ag}(\theta)$ = resistance of silver layer,

$R_{sol}(\theta)$ = resistance of solder (set to a large value for SuperPower wire),

$R_{hst}(\theta)$ = resistance of hastelloy or equivalent.

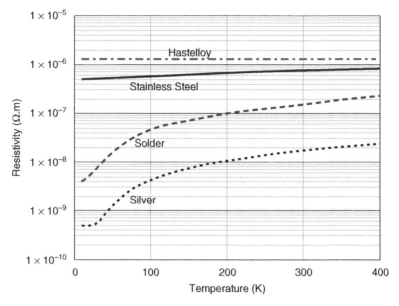

Figure 8.10   Resistivity of various metals as a function of temperature

These resistance values are functions of temperature dependent resistivity of each material shown in Figure 8.10. Resistance (R) of each element at temperature (θ) can be calculated with

$$R(\theta) = \frac{\rho(\theta)}{wt},\tag{8.10}$$

where

ρ = resistivity of material,

w = width of material,

t = thickness of material.

Once the effective resistance of the stabilized HTS strands is known, the total length of each strand (L_{strand}) necessary for limiting the fault current can be calculated from

$$L_{strand} = \frac{R_f N_{par}}{R_{eff}\left[(F_{oc}F_q I_o / N_{par}), T_{max}\right]},\tag{8.11}$$

where

R_f = required resistance of HTS FCL element,

R_{eff} = effective resistance of stabilized strand (Ω/m) at the maximum temperature (T_{max}) attained by HTS at end of fault hold-time,

N_{par} = total number of strands in parallel.

L_{strand} is also equal to the piece length of a CTC necessary for winding a coil or assembling in linear elements. The total length of strands needed for each phase is simply $N_{par} \times L_{strand}$.

Since a coil turn includes different materials with different resistivity and temperature dependences, it is best to calculate coil temperature with a step-by-step integration process. At any instant in time, incremental heat load (Q) generated in a conductor is given by

$$Q(I,\theta) = I^2 R(\theta)\delta t, \tag{8.12}$$

where

I = current in the conductor,

R = resistance per meter length of conductor,

δt = time integration step.

Incremental temperature rise of the conductor due to Q is given by

$$\Delta \text{Temp}(I,\theta) = \frac{Q(I,\theta)}{\begin{array}{c} V_{hts}\eta_{ss}(\theta) + V_{hst}\eta_{hst}(\theta) + V_{ss}\eta_{ss}(\theta) + \\ V_{ag}\eta_{ag}(\theta) + V_{sol}\eta_{sol}(\theta) \end{array}} \tag{8.13}$$

where

V_{hts}, η_{ss} = volume and heat capacity of HTS per metric length of the conductor,

V_{hst}, η_{hst} = volume and heat capacity of hastelloy per metric length of the conductor,

V_{ss}, η_{ss} = volume and heat capacity of stainless steel per metric length of the conductor,

V_{ag}, η_{ag} = volume and heat capacity of silver per metric length of the conductor,

V_{sol}, η_{sol} = volume and heat capacity of solder per metric length of the conductor.

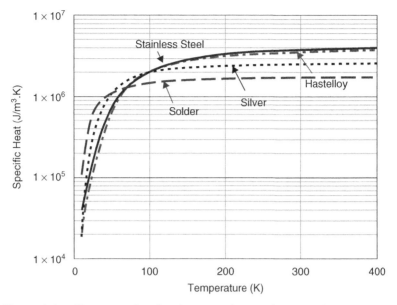

Figure 8.11   Heat capacity of various metals as a function of temperature

Heat capacity of various metals is shown in Figure 8.11 as a function of temperature.

During normal steady-state operation, the voltage across the HTS element is zero. Suppose that V_{ph} is the source voltage of the infinite bus, the voltage drop in the grid reactance (resistance is neglected) is $X_g I_o$, where X_g is the grid reactance and I_o is the phase current. The voltage drop across load is simply $X_l I_o$, where X_l is load reactance. The sum of $X_g I_o$ and $X_l I_o$ equals to V_{ph}. The instantaneous rate of change of line current (pI_t) is given by

$$pI_t = \frac{1}{(L_g + L_l)}\left\{V_{Iph} - L_{sh}\left[\frac{R_{fcl}}{L_{sh}}(I_t - I_{sh})\right]\right\}, \qquad (8.14)$$

where

pI_t $= \delta I_t/\delta t$ = rate of change of current,
L_g, L_l = inductances of grid and load, respectively,
V_{Iph} = instantaneous phase voltage,
L_{sh} = inductance of shunt reactor,
R_{fcl} = resistance of HTS FCL element.

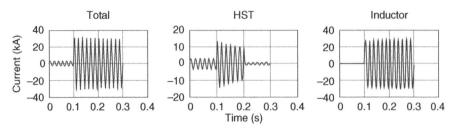

Figure 8.12 Total line, FCL, and inductor currents as a function of time

Equations (8.13) and (8.14) could be integrated simultaneously using a step-by-step integration routine. Load inductance (L_l) is set to zero for simulating a fault.

The integration process is performed with an initial guess of steel stabilizer thickness (t_{ss}). If temperature of HTS is higher than T_{max}, then t_{ss} is increased, and if it is less than T_{max}, then t_{ss} is reduced. With a few tries, the correct value of t_{ss} can be obtained. Integration is performed with a Runge–Kutta integration routine. Results (as a function of time) of the integration process are presented in Figure 8.12. The figure shows a variation of the total line current, HTS FCL element current, and shunt inductor current as a function of time. In each plot, the first 0.1-second period represents steady-state normal operation, the next 0.1-second period represents the fault (when the load reactance is set to zero), and the last 0.1-second period represents the point when the HTS FCL has been isolated (i.e., removed from the circuit). As expected, during the normal period, the line current is entirely carried by the HTS FCL element and the current in the shunt reactor is zero. During the fault period, the HTS FCL carries its design current (~10 kA) and the shunt inductor carries the balance of the current. Also during this fault period the HTS FCL current gradually decreases, whereas the shunt inductor current increases. Once the HTS FCL is isolated by opening the fast switch or circuit breaker (CB), the entire current is carried by the shunt inductor. The total current remains at <30 kA during the entire fault period, as required by the design specifications. The resistance of the HTS FCL and its temperature are shown in Figures 8.13 and 8.14. The final temperature attained by the HTS element is 300 K, as defined in the design specifications.

The SuperPower production wire must be reinforced with a 127-µm-thick stainless steel tape for use in the FCL. Since SuperPower does not have a process to laminate the thick stabilizer to their wire, it is necessary to develop alternate techniques. The design details of the FCL utilizing the SuperPower wire are summarized in Table 8.3. Based

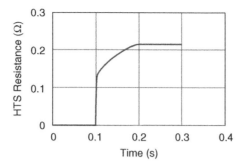

Figure 8.13 Resistance of HTS element as a function of time

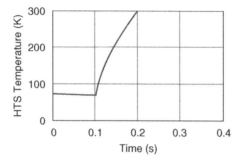

Figure 8.14 HTS FCL element temperature as a function of time

on the specifications in Table 8.2, the HTS FCL circuit has to generate a resistance of $0.22\,\Omega$ at the end of specified fault period of 0.1 second. The impedance of the inductor in parallel with the HTS FCL circuit is $0.073\,\Omega$. The FCL assembly is assumed to have three-phases connected in a Y-configuration. A critical current of $300\,\text{A/cm}$ (self-field, $77\,\text{K}$) is assumed for the 2G wire. The width of a 5×17 CTC is assumed to be $12\,\text{mm}$, and the turn-to-turn insulation width is $17\,\text{mm}$. The thickness of a CTC with a stabilizer is $2.4\,\text{mm}$ for SuperPower strands. The thickness of spacer between turns is $3\,\text{mm}$. The spacer is made of crinkled paper (currently used in conventional transformers) with a total built of $3\,\text{mm}$. The cross section of a bifilar pancake is shown in Figure 8.15. A distance of $20\,\text{mm}$ is maintained between adjacent coils during stacking of pancake next to each other. The minimum piece length of CTC used in a coil is about $45\,\text{m}$. The coils are of bifilar construction type wherein adjacent turns carry current in opposite direction, as shown in Figure 8.1. The maximum piece length of a CTC could be reduced to about $22.5\,\text{m}$ if two pieces of CTC are joined in the bore of a coil. The total

Table 8.3   Example FCL design details

FCL, 13.8 kV, 3 kA	AMSC	SuperPower
Line voltage, kV-rms	13.8	13.8
Norma current, kA-rms	3.0	3.0
Base impedance, ohm	2.656	2.656
System impedance, pu	0.075	0.075
System inductor impedance, pu	0.027	0.027
Voltage across HTS circuit during fault, pu	0.260	0.260
Unlimited fault current, kA	40	40
Limited fault current in the whole circuit, kA	29.3	29.3
Limited fault current in the HTS circuit, kA	9.6	9.6
Basis insulation level (BIL), kV	95	95
FCL impedance at the end of fault, ohm	0.22	0.22
Fault hold time, s	0.1	0.1
Operating temperature, K	72	72
Temperature at the end of fault, K	304	300
Wire critical current at 77 K (A/cm), A	150	150
Wire (strand) width, mm	5	5

Wire Layers

	AMSC	SuperPower
HTS layer thickness, μm	1	1
Hastelloy thickness, μm	75	100
Silver thickness, μm	2.5	2
Solder thickness, μm	20	0.1
Stainless steel stabilizer thickness, μm	120	127
Maximum temperature limit at end of fault period	300	300
Effective resisitance of non-HTS at T_{max}, ohm/m	0.074	0.079

CTC Cable Specification

	AMSC	SuperPower
Strand width, mm	5	5
Number of strands	17	17
Cable width, mm	12	12
Cable thickness, mm	2.32	2.43
Pancake coil ID, mm	200	200
Pancake coil OD, mm	604	591
Turns/pancake	38	36
CTC length in a pancake, m	48.0	44.7
Strand length/phase, km	3.3	3.0
Strand length/3-phase, km	9.8	9.1
Number of pancakes/ph	4	4
Width of a pancake coil, mm	17	17
Separation between pancakes, mm	20	20
Axial length of pancake stack/ph, mm	148	148
Voltage drop across a pancake, kV	1.04	1.04
Pancake in series/phase	2	2
Pancake in parallel/phase	2	2
FCL resistance at the end of fault, ohm	0.217	0.216

Shunt Current

	AMSC	SuperPower
Just before the HTS coil is disconnected, kA-rms	28	28
After HTS coil is disconnected, kA-rms	29	29
Energy dissipated/phase, MJ	2.38	2.34
Inside diameter of cryostat, m	1.32	1.30
Height of coil stack, m	1.04	1.04

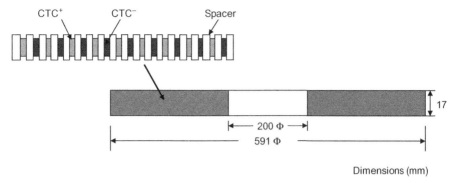

Figure 8.15 Cross section of a bifilar pancake

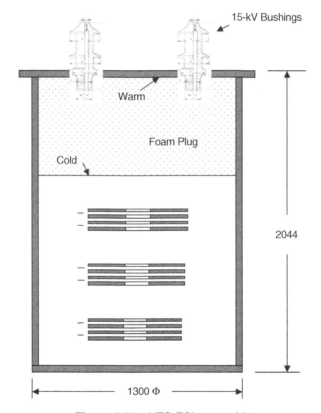

Figure 8.16 HTS FCL assembly

lengths of the CTC and the strands required for all three phases are roughly 540 m and 93 km, respectively.

A R-FCL assembly is shown in Figure 8.16. It houses the HTS FCL components for all three phases only. The external inductor is not included in this package. The diameter of the cryostat is 1300 mm, and

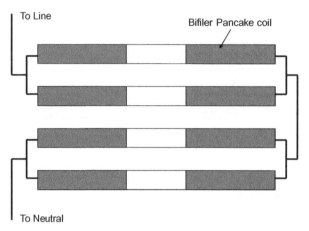

Figure 8.17 FCL HTS coil assembly for one phase

it is 2040 mm tall (including the space for the HV bushings.) The top of the cryostat has a tight fitting foam plug that separates the cold LN_2 region from the warm top flange. The top flange also houses six 15-kV bushings. These are off-the-shelf conventional bushings. The distance between the cold LN_2 region and the top flange is about 800 mm, which is considered sufficient to house current leads with one end at LN_2 temperature and other end at near room temperature. The LN_2 is subcooled using the cooling approach discussed in the Chapter 3.

Each phase assembly consists of four pancake coils: two in series and two in parallel. The connection arrangement is shown in Figure 8.17. A corona ring with a minor diameter of 31 mm surrounds each coil. The clear distance between the coronal ring and the inside wall of the cryostat is 300 mm, which is considered sufficient for withstanding BIL voltage of 385 kV. The coil arrangement discussed above will work fine for voltages up to the 15-kV class devices. More complex arrangements would be required for higher voltage applications.

The procedure described above for designing an FCL with SuperPower wire can also be used for designing with the AMSC wire. The design details for an FCL employing AMSC wire are also listed in Table 8.3.

A CTC built with 2G wire is a good choice for building fault current limiters. It reduces the number of individual coils and thereby reduces the total package size. The utilities are particularly sensitive to the overall size of an FCL assembly as it must fit into a restricted space in a substation. The most attractive application is at the subtransmission and transmission levels (>138 kV), where a CTC-based FCL would be most attractive. However, in this case the bushings are large and significant clearance is required within the cryostat, making the overall system large.

A point of special concern is the voltage withstand capability between the end turns of a bifilar coil, especially under conditions after a fault when the conductor is normal. A lightening strike at this time can apply a large voltage between those turns. Fortunately, it can be shown that the capacitance between a stack of bifilar coils will cause a BIL voltage to distribute across all the coils evenly. Even so, the voltage can be very substantial, and it may be necessary to have voltage arresters to further protect the system.

8.3.2 Example Design—Saturated Core FCL

A 3-Φ configuration of a saturated iron-core FCL is shown in Figure 8.8. Although a common HTS coil is employed for saturating all six cores for a 3-Φ system, it is possible to study and perform a preliminary design with the single phase model shown in Figure 8.18. This model consists of two copper iron-core coils, which are electrically connected, in series with the line being protected. The cores are driven into the saturation state with a common superconducting coil. The two copper coils are wound with opposite polarity such that at any given instant, the flux in one core demagnetizes (bucks) while the flux in the other core drives it further into saturation (boosts). The model in Figure 8.18

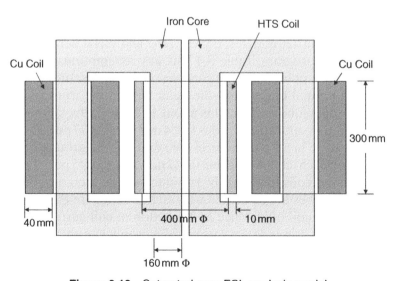

Figure 8.18 Saturated core FCL analysis model

is used for calculating the copper coil's inductance as a function of the iron-core saturation level and current in the copper coils. The iron core employs M-3 material for inductance calculations (also used by Zenergy Power [19]).

The FCL is sized to meet FCL requirements in Table 8.2. The system impedance (i.e., impedance between the infinite voltage source and the bus) is 0.2 Ω (or 0.075 pu) is calculated using the unlimited fault current (40 kA) specified in Table 8.2. Once the fault current is limited to 30 kA (as specified in the table) with the FCL, the total impedance required between the fault point and the infinite voltage (13.84 kV) source is 0.266 Ω (or 0.1 pu). The FCL must supply the additional needed impedance of 0.067 Ω (or 0.025 pu). The two copper coils connected in series, in the line being protected, supply this impedance. The net impedance experienced by the line is equal to the impedance of one coil with the iron core driven further into the saturation region and the second coil with the iron core driven toward the linear unsaturated state. When carrying the normal load current, copper coils induce a constant impedance that is equal to the sum of impedances of the two coils with the iron core in the saturated state. The effective copper coils' impedance increases by about 23% during a fault, once one of the cores transitions from the saturation region to the linear region. The analysis below estimates the copper coil impedances during normal and fault operations.

The example design model has each iron core with a 160-mm diameter. Each core is surrounded with a 300-mm-tall copper coil with an ID and OD of 180 and 260 mm, respectively, and each core has 14 turns. The superconducting coil has a single turn (assumed) and has an ID and OD of 400 and 420 mm, respectively. The analysis is performed with a finite-element program. Table 8.4 lists key assumptions. The superconducting coil supplies 800 kA turns of excitation to saturate the iron cores during normal operation—the peak field in the iron cores penetrating the superconducting coil is about 10 T. The inductance of each copper coil at this saturation level is 0.054 mH (or 0.0077 pu). However, if these coils were in air with no iron core and no background field, the inductance of each coil would be 0.017 mH (or 0.0245 pu). Thus the iron-core copper coils have nearly 3x inductance of their air-core state even when the cores are saturated with 10 T field. Figure 8.19 shows an iron-core copper coil inductance as a function of coil current. The coil inductance ratio (iron core/air core) is 3.24 even when the coil current is 100 kA and field in the iron core is 21.9 T. This shows that the inductance of the copper coils remains high even at very high saturation levels of the iron core.

Table 8.4   Key assumptions for Sat-FCL analysis

System bus line voltage, kV	13.8
Rated current, kA	3
Base pu impedance, ohm	2.656
Unlimited fault current, kA	40
Limited fault current, kA	30
System impedance, ohm	0.199
Infinite bus voltage, kV	13.84
Impedance supplied by FCL, ohm	0.067
System frequency, Hz	60
Current in the superconducting (SC) coil, kA	−800
Magnetic field in the iron core, T	−10.13
SC coil inductance with saturated core, μH	0.267
Copper coil details	
Number of turns	14
Coil inductance in air, mH	0.017
Coil inductance in air, pu	0.002

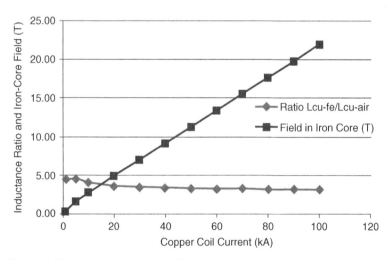

Figure 8.19   Iron-core copper coil inductance as a function of coil current

The inductance of the iron-core copper coil (in presence of saturation flux from the superconducting coil) is calculated using the following steps:

1. Calculate flux linkages with the copper coil using the specified superconducting coil excitation level.

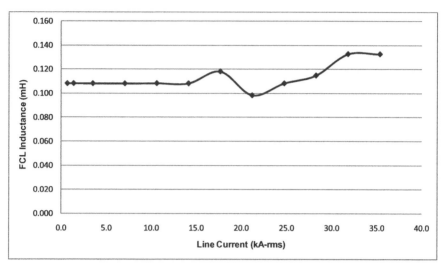

Figure 8.20 Inductance inserted in line as a function of normal AC coil current

2. Add a small current (to produce demagnetizing flux) in the copper coil and calculate total flux linkages with the copper coil.
3. The difference in flux linkages with the copper coil in steps 1 and 2 is that caused by the current in the copper coil. This difference in flux linkages divided by the copper coil current is the inductance of the copper coil at this copper coil current and iron-core saturation level.

The calculations above are repeated with small incremental steps from a small line current to the maximum 35 kA-rms fault current, and the copper coil inductance is calculated at each step. The total copper coil inductance as a function of the line current is shown in Figure 8.20. Once the iron core has been driven out of saturation with 45 kA-DC (corresponding to 32 kA-rms) fault current, the combined inductance of the two copper coils increases to 0.133 mH (or 0.019 pu). This increased induction in combination with the system impedance (0.075 pu) limits the fault current to 32 kA-rms. With these system and FCL impedances used at the 32 kA-rms fault current level, the voltage drop in the FCL copper coils is calculated to be only 0.2 pu (= 0.019/(0.019 + 0.075)).

The insertion impedance as a function of line current is shown in Figure 8.20. The insertion impedance remains constants until about 25 kA and then increases between 25 and 35 kA. The self-inductances of the copper and superconducting coils and their mutual inductances

Table 8.5   Inductances of copper and superconducting coils

Parameter	Iron Core Saturated	Iron Core Not Saturated
Copper coil self-inductance, mH	0.5787	0.0787
Superconducting coil self-inductance, μH	0.2669	0.3347
Mutual inductance between the coils, μH	3.427	4.615
Coupling coefficient between coils (k)	0.276	0.899

were calculated for both cases, that is, when the core is fully saturated and when it is in linear range. These values are listed in Table 8.5. During normal operation when the core is fully saturated, the coupling between the copper and superconducting coils is quite low ($\sim k = 0.276$) but is high ($k = 0.899$) when the core is not saturated.

The ripple current induced in the superconducting coil is proportional to the net flux linkages with it due to the fluxes in the two copper coil cores penetrating the superconducting coil. The net AC induced in the superconducting coil is given by

$$I_{sc} = \left(\frac{M_u}{L_{su}} - \frac{M_s}{L_{ss}} \right) I_p,$$ (8.15)

where

I_{sc} = ripple AC induced in the superconducting coil,

I_p = current in the copper coils,

L_{su}, M_u = superconducting coil self-inductance and mutual inductance with copper coils when the iron core is not saturated,

L_{ss}, M_s = superconducting coil self-inductance and mutual inductance with copper coils when the iron core is saturated.

With 3 kA-rms in the copper coils, the values of L_{su} and M_u are nearly equal to L_{ss} and M_s, and therefore the ripple current induced in the superconducting coil is nearly zero. However, with a 30-kA-rms fault current in the copper coils (during a single-phase, line-to-ground fault), the induced AC in the superconducting coil, as calculated using equation (8.15), is 28 kA or 3.5% ripple on 800-kA-DC in the superconducting coil. This ripple current will be smaller if the three-phase currents are nearly balanced. If the three-phase currents have an unbalance of 10%, then the AC ripple would be about 1/10th (0.35%). Nevertheless, the AC ripple current only lasts for the duration of the fault, and the superconducting coil can be designed to withstand any heat load due to the induced AC ripple.

The design procedure above for the saturated core FCL is for an illustration purpose only. There is lot of room to optimize the design for a specific application.

8.4 CHALLENGES

All HTS fault current limiters have their challenges when integrated in an electric grid. There are two types of R-FCLs:

1. An HTS coil in series with the line being protected.
2. A parallel combination of an HTS coil and inductor placed in series with the line being protected. The HTS coil switches the fault current to the inductor after a pre-determined period.

The challenges of resistive and inductive limiters are discussed below.

8.4.1   Challenges of Resistive FCL

The R-FCL shown in Figure 8.9 has a conventional inductor in parallel with the HTS FCL element. The purpose of the inductor is to limit the current to the desired level when the HTS FCL is not in the circuit. During normal operation the resistance of the HTS FCL element is zero, and therefore the reactance of the inductor is short-circuited. However, when the HTS FCL is not active, the system has to operate with the additional reactance of the inductor in series with the line. The impact of the this reactance on system performance should be studied prior to specifying this type of FCL. Before selecting this FCL option, the impact of added inductance on system performance should be evaluated. The added inductance may degrade the system performance by reducing power transfer capability during an allowable high current through fault transients. Normally the high current capability is essential for handling in-rush currents of transformers and large motors during starting, and for providing power transfer between buses connected through the high-impedance lines.

The fast circuit breaker (CB) in the HTS FCL loop is designed to open if the fault hold-time exceeds 0.1 second (or any other time interval assumed by a designer). However, in case the CB fails to open, the temperature of HTS element will keep rising and may permanently damage it. Thus backup CBs may be required to ensure high reliability. It is also possible to experience faults shorter than the hold-time but of sufficiently long duration to raise temperature of the HTS element above its critical temperature. In such events it would be necessary to detect the temperature of the HTS element and isolate it to allow recovery to its normal operating temperature. What fraction of time a FCL is active will depend on its location in a grid.

In the FCL design developed by SuperPower, the HTS element is permanently short-circuited with an inductor (made of normal conductor), so any losses occurring in the inductor must also be removed by the LN_2 cooling system. Recovery of such an FCL under load is very challenging. Any fault of duration shorter than the hold-time could trigger the HTS element to its normal state and transfer its current to the inductor. Under such situations both the HTS element and the inductor will be creating joule heating, so the cooling system must have to be of sufficiently large capacity to handle the thermal load. However, the thermal mass of copper inductor connected across the HTS elements initially absorbs the heat load adiabatically, much like "the steel tape laminated to HTS tapes" approach discussed in the example design of Section 8.3.1.

In all the resistive limiters the joule heating created during a fault generates a lot of bubbles that degrade the dielectric properties of the LN_2 and create a potential for a dielectric failure inside the FCL assembly. The HV insulation coordination of the R-FCL stack in the LN_2 environment is very critical and must be designed with utmost care. The many lessons learned from the HTS transformer's insulation development [20] could be used in designing R-FCLs. The space required to handle high voltages in the cryogenic environment makes these systems large.

8.4.2 Challenges of Inductive FCL

The inductive limiters discussed in Sections 8.2.2 and 8.2.3 also have their own challenges as described below.

Shield Iron-Core FCL   This FCL, shown in Figure 8.7, has challenges similar to those discussed for the R-FCL in Section 8.4.1. The FCL element (a hollow cylinder) made of a bulk HTS is prone to cracking because of the uneven heating of the wall, and therefore its long-term reliability in a grid is questionable. This FCL is similar to a transformer consisting of an iron core designed with the same flux density as that used in a conventional transformer. This makes the device bulky and heavy.

Saturated Iron-Core FCL   This FCL shown in Figure 8.8 might be more reliable. Although the superconducting coil used for saturating the iron cores does not experience the AC field created by copper coils in series with a line being protected during the normal operation or during a three-phase symmetrical faults, it has other challenges as discussed below. A finite AC ripple is created in the superconducting coil during a single phase to ground fault (and during line–line faults). The superconducting coil must be designed to handle any losses due to the ripple currents:

· The FCL could be significantly larger in size and weight than in other options.
· Significant inductance and resistance is added in the protected circuit even during normal operation

Like the R-FCL with an inductor in parallel, the impact of inductance added by the saturated core FCL on the system performance should be studied carefully. The added inductance may degrade the system performance. Minimization of the added inductance during normal operation is a significant challenge [21].

8.5 MANUFACTURING ISSUES

Resistive HTS FCLs based on 2G wire require wire availability in large quantity with consistent quality and performance, and at economically attractive price. Furthermore the 2G wire must be laminated to a high-resistivity stabilizer (e.g., stainless steel) to absorb joule heating adiabatically during a fault of short duration. The lamination process must create a good thermal and mechanical bond with the HTS wire, and should be suitable for use in a factory setting. Rapid progress is being made on all these requirements.

In the case of a bifilar coil, the maximum coil voltage is experienced between the two outermost turns next to the leads. Insulation on these turns has to be sufficient to withstand the BIL voltage. Since a significant portion of the BIL voltage is experienced by the first few turns or pancakes of a stack of coils, it is necessary to apply extra insulation on these turns. However, the extra insulation also impedes cooling. Thus the selection of an insulation scheme must be made by balancing insulation and cooling considerations.

In bulk material type FCL elements, the impact of an element failure and its impact on the FCL unit need to be evaluated beforehand. The manufacture of shatter-proof elements is challenging. However, the bulky shunt attached in parallel with the HTS contains shattered particles in the event of a broken element. No arcing is experienced, but the AC-losses increase significantly during normal operation. The broken element could be replaced later, if necessary. The first commercial units are based on such elements [11].

If FCLs are applied in remote locations, then their cooling system should have sufficient overhead in built redundancy to keep the FCL functioning with a failed cooling system while waiting for a repair crew.

8.6 PROTOTYPES

Two FCL projects are presently being developed with US-DOE support (an earlier program led by SuperPower has been canceled). American Superconductor Corporation in collaboration with Siemens and Nexans is developing a resistive HV FCL, and Zenergy Power is developing a saturated core FCL. Discussion below is based on a US-DOE study [22] for assessing testing requirements for various FCLs.

8.6.1 American Superconductor Corporation's (AMSC) Fault Current Limiter

American Superconductor Corporation has taken the lead to develop and demonstrate in-grid testing of a commercially viable three-phase transmission voltage superconducting FCL operating at 115 kV with Southern Cal Edison. Figure 8.21 shows the conceptual arrangement of pancake coils. Phase 1 of the project involves development of the core technology followed by a demonstration of a single phase FCL in the beginning of 2010. Phase 2 of the project will include the construction, test and in-grid operation of a full three-phase 115-kV FCL by the end of 2012.

American Superconductor Corporation has already conducted testing with their partner, Siemens, on a single-phase device with a rated current of 300 A-rms and a rated voltage of 7.6 kV, which corresponds to a nominal apparent power of 2.25 MVA. The testing was

Figure 8.21 American Superconductor Corporation's FCL concept (Courtesy AMSC)

completed in January 2007 at the IPH-Berlin test facility. This module corresponds to a 13-kV class three-phase module. The test demonstrated that the module could reduce a short-circuit current from 28 to 3 kA. American Superconductor Corporation and Siemens conducted R&D testing on the FCL module to validate its design and provide data for scaling up to a higher voltage class. To reduce the number of coils and hence the system cost, AMSC has developed a wider 12-mm stainless-stabilized wire. Siemens has wound this wire into bifilar coils and has demonstrated a successful single-phase test at distribution voltage.

At this time the utility-testing requirements for the full-scale FCL are still under development. The device rated at 138 kV will operate at 115 kV in the Southern California Edison (SCE) territory, due to an absence of 138-kV substations. The design criteria for the device is to reduce a fault from 63 to 40 kA. The design is also modular so that

Figure 8.22 SuperPower's FCL concept (Courtesy SuperPower)

pancake coils may be added or removed in series and in parallel configurations. This way the design may be extended to virtually any steady-state current or limiting requirement. In addition, by employing an external reactor, some flexibility is retained even in an existing installation to respond to system growth or change. This FCL is planned to be tested in accordance with IEEE and IEC specifications for 138-kV rated cable accessories and transformers.

8.6.2 SuperPower's Fault Current Limiter

SuperPower was developing a superconducting FCL for operation at 138 kV. The device utilizes a matrix design consisting of parallel 2G HTS elements and conventional shunt coils as shown in Figure 8.22. The program included the fabrication and testing of three prototypes: a single-phase proof-of-concept prototype, a single-phase alpha prototype, and a three-phase beta prototype. The first prototype unit was tested at KEMA Power Test Facility in Philadelphia, Pennsylvania, and the second prototype was planned to be tested off grid. The final beta prototype was to be installed and operated in the American Electric Power (AEP) grid. SuperPower tested two alpha prototype single-phase modules from 100 to 400 volts with a 1.2-kA-rms current and a 37-kA peak fault capability. They successfully proved the concept of recovery under load for AEP's reclosure sequence. SuperPower was optimizing the design to make it is more compact while still having the same functionality. Their final design intended to reduce fault currents by 20% to 50%. However, the company has terminated the program for unexplained reasons.

8.6.3   Zenergy Power's Fault Current Limiter

Zenergy Power is leading the design, constructions, and testing of a saturable iron-core superconducting FCL. It is a prototype for a commercial product suitable for operation in a typical 138-kV transmission grid substation. One of these devices will operate at distribution voltage (less than 69 kV) and another will operate at a transmission voltage of at least 138 kV. Zenergy's FCL prototype completed its first R&D tests at 480 V and 460 A in October 2007 at Pacific Gas and Electric (San Ramon, California). They also tested a three-phase 13.1-kV device at 10- and 16-kA fault levels at the PowerTech Laboratory in British Columbia, Canada, in December 2007. The device was able to reduce the 39-kA perspective peak current to 23 kA.

During 2008 Zenergy Power built a second full-scale, distribution-voltage FCL, culminating in testing at PowerTech Laboratories in October 2008. This device (Figure 8.23), which is rated for a nominal 1200-ampere steady-state, 15-kV class, was built in cooperation with the CEC and incorporated several improvements compared to the device of a year earlier. The October 2008 testing demonstrated a fault current

Figure 8.23   Zenergy's FCL concept (Courtesy Zenergy)

reduction of approximately 30%. After completing the tests at Powertech, the full-scale FCL was installed in the Avanti "Circuit of the Future" at Southern California Edison's (SCE) Shandin substation in San Bernardino, California. The "Circuit of the Future" is a DOE, CEC, SCE cooperative effort to showcase new technologies for the "smart" electric power grid. The Zenergy Power FCL was energized in the Avanti circuit on March 6, 2009, becoming the first HTS FCL to be operating in the US electricity grid.

Zenergy Power is designing a medium voltage device to be built and tested with a utility partner by mid 2010. The device is being designed to reduce a prospective fault current by 50%. Their final 138-kV device will reduce a 60- to 80-kA fault by 20% to 40%.

8.6.4 Nexans's Fault Current Limiter

This project was sponsored by the German Ministry of Education and Research. The CURL10 FCL is a 10-kV, 10-MVA device [23] with a continuous current rating of 600 A. It was built by Accel using BSCCO-2212 bifilar rods from Nexans. The device shown in Figure 8.24 was the first field test of a resistive HTS FCL and was installed in Germany's RWE Energie utility grid in 2004. It underwent a series of tests. In the

Figure 8.24 CURL-10 fault current limiter (Courtesy Nexans Superconductors)

Figure 8.25 FCL supplied to Applied Superconductor Ltd., UK (Courtesy Nexans Superconductors)

laboratory the resistive HTS FCL limited a prospective short-circuit current from 18 to 7.2 kA. It operated in the electric grid for nine months, and while it experienced several lesser faults, it did not experience a design fault. A design fault is the maximum fault level the device can limit given its internal characteristics. This is of significance because it shows that even if a device is placed in a real-world scenario, it may not undergo the worst-case scenario fault during the test period. This means that utilities need additional data to prove that the device functions properly. Nexans has supplied a FCL unit (Figure 8.25) for application in 12-kV grid with nominal current of 100 A. The FCL [24] limits the first peak to <6 kA in a system with perspective current of 22 kA.

8.7 SUMMARY

Most of the existing T&D infrastructure in the United States is reaching the end of its useful life and, coupled with the steady growth in electricity demand and the addition of a new generation, is increasing the fault current levels well beyond the capability of existing equipment. Utilities always look for ways to get more out of their existing equipment. The HTS FCLs present an option to rein in the fault current levels to within the capability of existing equipment. To help address these problems, with R&D funding from the US Department of Energy, equipment manufacturers, electric utilities, and researchers from private industry, universities, and national laboratories are teaming up to spur innovation and development of new technologies, tools, and techniques. Because of these efforts, the future electric grid will likely incorporate technologies very different from those that have been traditionally employed. The S-FCL is one of these technologies, and the first units are already being deployed commercially. Manufacturers and users are already working on developing standards for FCLs under IEEE.

REFERENCES

1. E. Thuries, V. D. Pham, Y. Laumond, T. Verhaege, A. Fevrier, M. Collet, and M. Bekhaled, "Towards the Superconducting Fault Current limiter," *IEEE Trans. Power Delivery* 6(2): 801–808, 1991.

2. D. Ito, E. S. Yoneda, K. Tsurunaga, T. Tada, T. Hara, T. Ohkuma, and T. Yamamoto, "6.6 kV/1.5 kV-Class Superconducting Fault Current Limiter Development," *IEEE Trans. Magnetics* 28(1): 438–441, 1992.

3. P. Tixador, Y. Brunet, J. Leveque, and V. D. Pham, "Hybrid Superconducting a.c. Fault Current Limiter Principle and Previous Studies," *IEEE Trans. Magnetics* 28(1): 446–449, 1992.

4. S. S. Kalsi, and A. Malozemoff, "HTS Fault Current Limiter Concept," 2004 *IEEE Power Engineering Society General Meeting*, June 6–10, 2004, Vol. 2, 1426–1430, DOI 10.1109/PES.2004.1373103.

5. L. Salasoo, K. G. Herd, E. T. Laskaris, and M. V. K. Chari, "Comparison of Superconducting Fault Current Limiter Concepts in Electric Utility Applications," *IEEE Trans. Appl. Superconductivity* 5(2): 1079–1082, 1995.

6. C. Weber and J. Bock, "Superconducting Fault Current Limiter SPI Project," The Power Delivery Applications of Superconductivity Task Force, *New York*, October 23–24, 2003, EPRI, Palo Alto CA.

7. R. Kreutz, J. Bock, F. Breuer, K.-P. Juengst, M. Kleimaier, H.-U. Klein, D. Krischel, M. Noe, R. Steingass, and K.-H. Weck, "System Technology and Test of CURL 10, a 10 kV, 10 MVA Resistive High-Tc Superconducting Fault Current Limiter," *IEEE Trans. Appl. Superconductivity* 15(2, Pt. 2): 1961–1964, 2005. DOI 10.1109/TASC.2005.849345.

8. J. Bock, F. Breuer, H. Walter, S. Elschner, M. Kleimaier, R. Kreutz, and M. Noe, "CURL 10: Development and Field-Test of a 10 kV/10 MVA Resistive Current Limiter Based on Bulk MCP-BSCCO 2212," *IEEE*

Trans. Appl. Superconductivity 15(2, Pt. 2): 1955–1960, 2005 DOI 10.1109/TASC.2005.849345.

9. W. Paul, M. Chen, M. Lakner, J. Rhyner, D. Braun, and W. Lanz, "Tl-based Patterned Superconducting Structures: Fabrication & Study," *Physica* C354: 27, 2001.

10. Office of Electricity Delivery and Energy Reliability, *Fault Current Limiters Bulletin of US Department of Energy*, March 3, 2009.

11. D. Klaus, A. Wilson, R. Dommerque, J. Bock, A. Creighton, D. Jones, and J. McWilliam, "Fault Limiting Technology Trials In Distribution Networks," Presented at CIRED, Prague, June 8–11, 2009.

12. T. Hori, A. Otani, K. Kaiho, I. Yamaguchi, M. Morita, and S. Yanabu, "Study of Superconducting Fault current Limiter Using Vacuum Interrupter Driven by Electromagnetic Repulsion Force for Commutating Switch," *IEEE Trans. Appl. Superconductivity* 16(4): 1999–2004, 2006. DOI 10.1109/TASC.2006.881809.

13. M. Endo, T. Hori, K. Koyama, K. Kaiho, S. Yanabu, K. Arai, and I. Yamaguchi, "Operating Characteristic of Superconducting Fault Current Limiter Using 24 kV Vacuum Interrupter Driven by Electromagnetic Repulsion," *Electric Power Applications, IET* (July): 363–370, 2009.

14. H. Saitoh, H. Ichikawa, A. Nishijima, Y. Matsui, M. Sakaki, M. Honma, and H. Okubo, "Research and Development on 145 kV/40 kA One Break Vacuum Circuit Breaker," IEEE Transmission and Distribution Conference & Exhibition, Asia Pacific, Vol. 2, October 2002, 1465–1468.

15. T. Tsutsumi, Y. Kanai, N. Okabe, E. Kaneko, T. Kamikawaji, and M. Homma, "Dyncmic Characteristics of High-Speed Opearted, Long Stroke Bellows for Vacuum Interrupters," *IEEE Trans. Power Delivery* 8(1): 163–167, 1993.

16. S. B. Abbott, D. A. Robinson, S. Perera, F. A. Darmann, C. J. Hawley, and T. P. Beales, "Simulation of HTS Saturable Core-Type FCLs for MV Distribution Systems," *IEEE Trans. Power Delivery* 21(2): 1013, 2006.

17. D. Verebelyi, J. Scudiere, A. Otto, U. Schoop, C. Thieme, M. Rupich, and A. Malozemoff, "Practical Neutral-Axis Conductor Geometries for Coated Conductor Composite Wire," *Superconduct. Sci. Technol.* 16: 1158–1161, 2003.

18. C. Weber, J.-C. Llanbes, D. Hazelton, and I. Sauers, "Transmission level HTS FCL," 2008 US-DOE Peer Review.

19. L. Masur, F. Darmann, and Lombaerde, "Design, Test and Demonstration of Saturable-Core Reactor HTS Fault Current Limiter," 2008 US-DOE Peer Review.

20. B. Schwenterly and E. Pleva, "HTS Transformer Development," Presented by Waukesha at the 2008 US-DOE Peer Review.

21. K. Smedley, "Development of Fault Current Controller Technology," California Energy Commission, Transmission Research Program Colloquium, Sacramento, September 11, 2008.

22. B. Marchionini, N. K. Fall, and M. Steurer, "An Assessment of Fault Current Limiter Testing Requirements," Prepared by Energetics Inc. for US-DOE, February 2009.

23. M. Noe and M. Steurer, "HTS Fault Current Limiters: Concept, Applications, and Development Status," *Superconduct. Sci. Technol.* (20): R15–R29, 2007.

24. Nexans' presentation in Nagoya, May 2009.

9

POWER CABLES

9.1 INTRODUCTION

An aging and inadequate power grid is now widely seen as the greatest obstacle to restructuring power markets in the United States. Utilities and users face several converging pressures brought on by combination of steady load growth, stringent barriers to siting new facilities, introduction of new competitive forces, and customer demand for improved power quality and reliability. The deregulation of transmission and distribution of electricity and the higher power quality requirements of a growing digital economy are encouraging utilities to react quickly to changing market conditions and offer new solutions to their customers. These considerations, in addition to the delay and cost complications associated with acquiring new rights-of-way, point to the need for new technologies that can increase the electrical capacity and flexibility of the electric grid by replacing the existing cables or overhead lines. The high temperature superconductor (HTS) cable is a promising new technology to address these issues. Listed below are the key benefits of employing HTS cables:

- Current-carrying capability three to five times that of a conventional cable.

Applications of High Temperature Superconductors to Electric Power Equipment, by Swarn Singh Kalsi

- Minimum waste heat or electrical losses and no soil heating.
- Possible installation in existing conduit infrastructure for breaking urban power bottlenecks.
- Less space than conventional cable, leaving room for new generation and load growth.
- High-power capacity at lower voltage enabling elimination of one or more transformers and associated ancillary equipment.
- Improved cost-effective control of power flow across meshed grids.
- Operating life of existing highly loaded infrastructure extended by taking some of the load.
- Use of environmentally benign LN_2 for cooling.
- No electromagnetic stray field emissions.
- Lower impedance [1] than conventional cables and overhead lines.

Since an HTS cable has characteristically lower impedance than an overhead transmission line or a conventional underground cable, more power could be transferred between the two points it connects. With the rapid decline in available underground space for conventional cables and the prohibitive cost of real estate for new or expanded substations, HTS cables are emerging as a solution for electric power in densely populated areas such as New York City. The reliability of electric power is enhanced by interconnecting substations and by relaxing substation space constraints.

Over the past decade several HTS cable designs have been developed and demonstrated all over the world [2,3,4,5]. All HTS cables cooled with LN_2 can be grouped into the following three broad categories:

- Single-phase cable with a dielectric at LN_2 temperature.
- Single-phase cable with a dielectric at room temperature.
- Triax™ cable* containing all three-phases in a single cable.

Each cold dielectric cable can be housed in an individual cryostat, or a set of three-phase cables could share a common cryostat. Both approaches have been employed in the United States and abroad. Currently medium- and high-voltage[†] cables are in operation in US electric grids. These cables are identified as follows:

---

*The Triax™ (trademark of Southwire Company) cold dielectric cable uses the same amount of HTS per phase as the warm dielectric because no HTS shield layer is required.
[†] General voltage classifications in the United States are: low = <5 kV, medium = 15–46 kV, HV = 69–230 kV, EHV = >230 kV.

- *High-voltage cable* [6]. 138-kV, 2500A built by AMSC and Nexans Cables is currently operating on Long Island, New York. One coaxial cable with an HTS shield for each phase is housed in an individual cryostat.
- *Medium-voltage cable* [7]. 34.5-kV, 800A cable built by SuperPower and Sumitomo Electric is operating in Albany, New York. All cables for the three phases share a common cryostat.
- *Medium-voltage cable* [8]. 13.2-kV, 3000A cable built by Southwire and nkt Cables is operating in Columbus, Ohio. This Triax cable has all three-phases incorporated in a single cable housed in a single cryostat.

Prior to the current HTS cable programs, high-capacity power cables were developed in the 1970s. Two examples are an NbTi cable [9,10] cooled with LHe and a resistive aluminum cable cooled with LN_2 (also called Cryocable [11,12]). However, these cables were not employed in the electric grid because by the late 1970s the demand for electricity had slackened. Moreover utilities were reluctant to employ a high-capacity link in their grid due to concerns that unplanned outages of the high-capacity link could trigger cascading blackouts. Even today the HTS cables are economical only when they can carry large current (i.e., high capacity). Also, utilities are still concerned about employing high-capacity cables, whose abrupt failure could destabilize a grid.

Moreover HTS cables are prone to quenching[‡] during high-current (40–80-kA) faults. Once a cable warms up following a quench, it takes considerable time to cool it back to its pre-fault operating temperature—thus contingency plans are needed to keep the grid stable while the cable is being cooled down. More HTS tapes (i.e., conductors) in a cable could carry higher fault current, but then the cost of the cable goes up due to the very high cost of HTSs. The HTS cables capable of limiting the fault currents are also envisaged, but the issue of cooling them back to their normal temperature in the shortest possible time remains.

Then again, cryocables are capable of withstanding high current faults like conventional cables, and no downtime is required while they are being cooled down to their pre-fault operating temperature. These cables offer up to three times the power transmission capability in the same space as a conventional cable and present a low capital cost option at the expense of an operating cost similar to that of a

---

[‡]As explained in Chapter 2, in a quench state a superconductor element becomes a normal resistive element, and its temperature rises rapidly due to I^2R losses caused by the current in it.

conventional cable of comparable rating. This cable could also carry partial loads when it is above its rated operating temperature.

The following section describes the configuration, design procedure, manufacturing, and operational challenges of all cables.

9.2   CONFIGURATIONS

Four cable configurations are discussed in this section: a resistive cryogenic cable and three HTS cables.

9.2.1   Resistive Cryogenic Cable

The resistive cryogenic cable (cryocable) employs an aluminum conductor cooled to the LN_2 temperature. Aluminum is preferred over copper because at the LN_2 temperature the resistivity of copper is only 30% lower than that of aluminum but aluminum is much cheaper and lighter. Because the technical feasibility of a cable system depends greatly on insulation performance under high-voltage conditions at a low temperature, the dielectric properties of various insulation materials were measured as small cable samples in liquid nitrogen [13]. Figure 9.1 shows a prototype system consisting of a set of three-phase cables in a single cryostat. A hollow bore permits the internal conductor's cooling of the cable. Each cable comprises a hollow former supporting

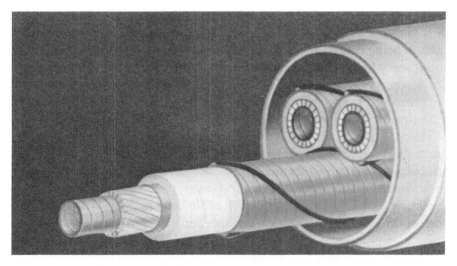

Figure 9.1   Resistive cryogenic cable configuration

a helically wound, stranded, transposed conductor made of aluminum (AL6061-T6). The conductor is lapped with a suitable electrical insulation impregnated with liquid nitrogen. The manufacturing processes of conductor and insulation system are the same as for conventional cables. A large refrigerator system based on turbo machinery was specified that requires 8 W of refrigerator input power to remove 1 W of cable loss. In the 1970s a full-size cable (500-kV, 1200-A) prototype was successfully built and tested. Concepts for high-voltage terminations and splices were also developed and tested. The resistive cryogenic cables are much lower in capital cost than superconducting cables, but they have a higher operating cost due to need for removing higher conductor loss using the refrigerator.

GE and Public Service Electric and Gas (PSE&G) of New Jersey jointly conducted economic feasibility of this system for a 30-mile application under a US Department of Energy contract [12]. The study concluded that a single cryocable could carry the same 1000-MW power as three oil-cooled conventional cables. Then capital costs of both options were similar if the three conventional cable installation were to be time-phased over a 10-year period. Although the conventional solution required more space than the cryogenic cable, the conventional solution offered higher system reliability because one conventional cable failure resulted in the loss of a third capacity compared to 100% loss of capacity for the cryocable. Even today, utilities are leery of large-capacity links due the potentially adverse impact on grid stability. However, both resistive cryocable and HTS cables are only economical for long-length high-capacity applications. The utility sector's concern about sudden loss of a high-power link is inhibiting the adaptation of new cable technologies like cryocables and modern HTS cables.

9.2.2 HTS Cable

A variety of cable designs have been prototyped and developed to take advantage of the efficiency and operational benefits of superconductivity, while minimizing the capital and operating costs due to the high cost of the HTS wire and refrigeration system. The variations in cable architecture have important implications in terms of efficiency, stray electromagnetic field (EMF) generation, and reactive power characteristics. At present there are two principal type of HTS cables. The simpler design has a single conductor, consisting of HTS wires stranded around a flexible core. This cable design (Figure 9.2) employs an outer dielectric insulation layer at room temperature and is called a "warm dielectric" design. The cable subassembly, consisting of HTS layers wrapped

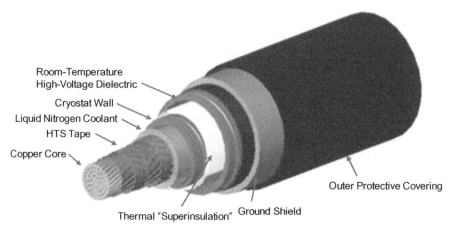

Room-Temperature
High-Voltage Dielectric

Cryostat Wall

Liquid Nitrogen Coolant

HTS Tape

Copper Core

Outer Protective Covering

Thermal "Superinsulation"   Ground Shield

Figure 9.2   Single-phase warm dielectric cable configuration

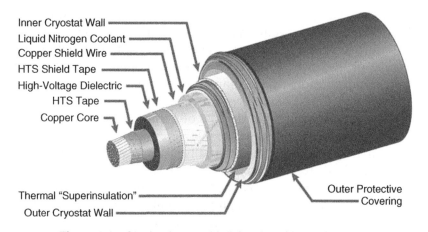

Inner Cryostat Wall

Liquid Nitrogen Coolant

Copper Shield Wire

HTS Shield Tape

High-Voltage Dielectric

HTS Tape

Copper Core

Thermal "Superinsulation"

Outer Cryostat Wall

Outer Protective
Covering

Figure 9.3   Single-phase cold dielectric cable configuration

around the inner flexible copper core, is first contained within a thermal insulating layer (cryostat). The electrical insulation is applied over the outer (room-temperature) wall of the cryostat. It offers high power density and uses the least amount of HTS wire (like a Triax cable) for a given level of power transfer. The drawbacks of this design relative to other superconductor cable designs include higher electrical losses (and therefore a requirement for more refrigeration power), higher inductance, and a required phase separation for reducing eddy-current heating and production of stray electromagnetic fields (EMF) in the vicinity of the cable.

Another option is a cold dielectric cable in which HTS wires are stranded around a flexible copper core (Figure 9.3). An electrical

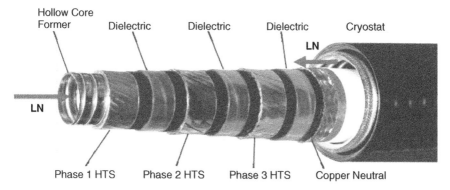

Hollow Core Former   Dielectric   Dielectric   Dielectric   Cryostat

LN

LN

Phase 1 HTS   Phase 2 HTS   Phase 3 HTS   Copper Neutral

Figure 9.4   Three-phase "Triax" cold dielectric cable

insulation is applied over the HTS layer, and a second layer of HTS tape is applied over the insulation to act as a cold shield for the single-phase cable. Liquid nitrogen coolant flows over and between both layers of HTS wire, providing both cooling and dielectric insulation between the center conductor layer and the outer shield layer. This is commonly referred to as a coaxial, "cold dielectric" design. This option uses the maximum amount of HTS wire but offers several important advantages, including higher current-carrying capacity, reduced AC losses, low inductance, and the complete suppression of any stray electromagnetic field (EMF) outside of the cable assembly. The reduction of AC losses enables wider spacing of the cooling stations and the auxiliary power equipment required to ensure their reliable operation. The inductance of a cold dielectric cable is also significantly lower than that of a conventional cable. Since the power flow in a circuit is inversely proportional to its impedance, the HTS cable could carry more load.

Another configuration of importance is the "Triax™" design where all three-phases are contained in a single cable as shown in Figure 9.4. This cable also has the lowest impedance and uses the least amount of HTS wire of all cold dielectric superconducting cable options.

The three-phases in a single cable is constructed with three electrically insulated layers of HTS wire built around a hollow former. This design reduces the AC losses, which means a lower refrigeration load. The concentric phases also lead to the elimination of stray electromagnetic fields. In addition this design requires only about half the superconducting materials required by the coaxial cold dielectric design and takes up less space, since a single cable is equal to three separate cables. The entire cable assembly is insulated and jacketed to protect it from thermal and physical damage. The cold LN_2, passing through the hollow central former along the length of the cable, cools it. The warm LN_2 is returned through a gap between the outer layer of the cable insulation and the inside wall of the cryostat.

9.3 DESIGN ANALYSIS

This section describes preliminary analysis for designing resistive cryocables and HTS cables. The analysis includes designs for the conductor, insulation, and cryostat. Readers can explore a specific cable design in detail by consulting the provided references.

9.3.1 Cryogenic Cable Analysis

A cryogenic cable consists of an aluminum conductor, dielectric insulation, and a cryostat to house a set of three cables in a single cryostat, as shown in Figure 9.5. Figure 9.6 shows an individual cable concept. The conductor bundles are wrapped around a hollow former. The cross section of a cryocable consists of a hollow aluminum stranded conductor, dielectric insulation over the conductor, a ground shield over the insulation, and skid wires over the ground shield. The skid wires are applied for reducing friction while cables are pulled through the pipe during installation. Analyses for designing various parts of the cable are discussed below. The cable splice and termination design concepts are similar to those employed for conventional cables.

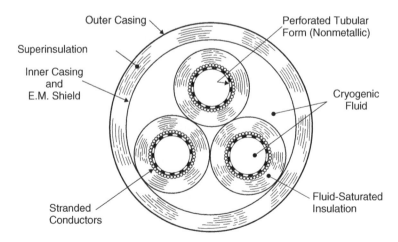

Figure 9.5 Cryogenic cable system cooled with LN_2

Former   Bundled Conductor   Insulation   Ground Shield

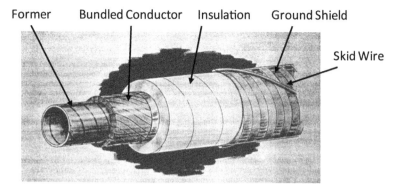

Skid Wire

Figure 9.6   A LN$_2$ cooled resistive cable

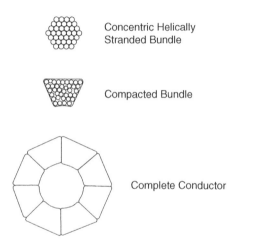

Concentric Helically
Stranded Bundle

Compacted Bundle

Complete Conductor

Figure 9.7   Concentric helically stranded compacted cable conductor

Conductor Design   The conductors for power cables of large cross section are commonly built up from concentrically stranded bundles, which may be compacted to trapezoidal or sector shape. The concentric stranding does not constitute perfect transposition relative to the non-uniform self-field of the cable. While coupling losses due to imperfect coupling may not be significant at room temperature, they can be large at the LN$_2$ temperature. Figure 9.7 shows the cross section of a typical conductor. A method for calculating the AC/DC ratio of a concentrically stranded, compacted bundle is described elsewhere [14]. A typical stranded and compacted conductor has AC to DC resistance ratio of around 1.3.

Insulation Design   A cable conductor is lapped with a dielectric tape, usually about 1-inch wide. A gap of about 0.05 to 0.1 inch is maintained between the adjacent tapes. When a cable is reeled, the tapes in successive layers slide and prevent wrinkling of tapes on the inside radius of the bent cable. A tape in an outer layer bridges the gap with the inner layer with a typical 50/50 overlap. Typical tape thickness is about 0.004 to 0.008 inch. In operation, the liquid nitrogen fills gaps among and between the tapes, and the dielectric strength of the liquid nitrogen filled gaps is usually comparable to that of the tape. Measurements are normally made on small cable samples where the insulation thickness consists of many layers of dielectric tapes applied in a manner similar to that of a final cable. Many dielectric materials were tested under a US-DOE program [15] between 1972 and 1974 using 10-ft-long samples with a 3-ft test region. The test samples had up to 200-kV, 60-Hz test voltages and impulse voltages applied to them. The breakdown strength of the cellulose/polypropylene paper laminate was determined to be 201 kV in a 0.22-inch-thick sample at a test length of 36 inches. The average and peak breakdown strength of this sample was 913 and 1099 kV/inch. Measurements made on different built-up insulation thicknesses were used to develop the scaling function given in equation (9.1). The peak strength of this material was projected to be around 900 kV/inch (36 kV/mm) for a cable insulation thicknesses of about 0.8 inch (=20 mm). From the measured data [15] constants A and B in equation (9.1) are estimated to be 47.8 kV/mm and 0.1, respectively. This equation is valid only for insulation thickness less than 50 mm:

$$E_m = A t_{ins}^{-B},\qquad(9.1)$$

where

E_m = peak electric stress (kV/mm),
t_{ins} = insulation thickness (mm),
A = 47.8 kV/mm,
B = 0.1.

Figure 9.8 shows a generic cable cross section with electric insulation. The insulation thickness for a cable can be determined based on a selected maximum working stress. The peak stress (E_m) normally occurs at the surface of the cable conductor and is given by

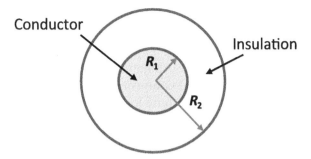

Figure 9.8 Generic cross section of a cable

$$E_m = \frac{V}{R_1 \ln(R_2 / R_1)}, \tag{9.2}$$

where

E_m = peak dielectric stress at conductor surface,
R_1 = inside radius of insulation (next to the conductor),
R_2 = outside radius of insulation built,
V = conductor voltage.

The equation above can be re-arranged to provide the radial thickness (t_{ins}) of the insulation:

$$t_{ins} = R_1 \left(e^{V/(E_m \cdot R_1)} - 1\right). \tag{9.3}$$

Generally, the peak stress (E_m) is selected as half to a third of the breakdown strength of the dielectric material. Thus for the polypropylene-cellulose paper laminate, a working stress of about 500 V/mil (= 20 kV/mm) is considered reasonable.

Usually the outside surface of a cable conductor is nonniform, and it is customary to cover it with a few layers of a carbon-black paper (i.e., semiconductive tape made from cellulose or polyethylene paper embedded with carbon particles). Similarly carbon-black layers are applied on the outside surface of the cable insulation prior to application of copper ground shield.

Dielectric materials operating in the high electric stress environment of power cables also generate significant thermal load (loss) that must be removed by the refrigerator. This loss component is called "tanδ loss." The tanδ losses for a variety of materials were measured [13] while the materials were impregnated with LN_2. Losses were minimum for polyethylene paper (DuPont "Tyvek") and highest for cellulose paper. However, the polyethylene paper had lower dielectric breakdown strength than the cellulose paper. The polypropylene-cellulose paper laminate provides a good compromise between the high breakdown strength and the moderate tanδ loss. For this reason this material has been widely used in conventional oil-impregnated cables since 1970 and has been selected by most LN_2 cooled resistive cryogenic and HTS cable designers.

Cryostat Design   A set of three-phase cables sharing a common cryostat is shown in Figure 9.5 (one phase per cable.) The cryostat consists of an inner cold wall and outer warm wall. Space between the two walls in filled with superthermal insulation (MLI). Although the balanced three-phase currents in the three cables produce essentially a zero leakage field a few meters away from the cable, at the inside radius of the cryostat the total field due to the three cables is not zero. This field has the frequency of the line current and generates eddy-current heating in the cryostat walls. To reduce the eddy-current heating, it is preferable to employ aluminum for the inner pipe of the cryostat. If a steel pipe is employed as the inner pipe, then it should be shielded with a layer of aluminum (~5 mm thick). The inner wall of the cryostat is designed to withstand a positive pressure of about 16 bar.

Listed below are the major components of thermal load of a cryogenic cable.

1. Resistive losses in the cable conductor (a function of conductor's cross section and current).
2. tanδ loss in the dielectric insulation.
3. Thermal conduction through cryostat walls (from the room-temperature environment to the LN_2 environment). This is generally a function of quantity of the MLI used and the type of mechanical support between the cold and warm pipes; in a good design it is possible to achieve a loss (thermal heat load) of 8 W/m length of cryostat with inner pipe diameter of 18 inch (450 mm).
4. Eddy-current losses in aluminum shield (a function of line current).
5. LN_2 pumping loss.

The sum of these loss components equals the needed refrigerator capacity. The separation between adjacent refrigeration stations is determined by inlet and outlet temperatures of the LN_2 and its flow velocity.

It is also possible to employ a nonmetallic pipe [12] to carry LN_2. The pipe could be insulated with 8-inch (200-mm)-thick foam insulation. In such a system the aluminum eddy-current shield is also eliminated. However, the heat leak through the foam insulation is higher than through the MLI. Any saving in the heat load achieved by elimination of the aluminum shield is erased by the higher heat leak through the foam insulation. In the final analysis, the comparison is between costs of vacuum insulated metallic pipes versus foam insulated nonmetallic pipe.

9.3.2   HTS Cable Analysis

This cable analysis is mostly aimed to construct of high voltage conductor and shield layers using HTS tapes. Since each HTS tape is about 4 mm wide, many tapes are laid helically around a central former in several layers to carry the rated and overload currents specified for a design. In the absence of a well-controlled design, all HTS tapes are not likely to share the current evenly. This problem is similar to that discussed for conventional stranded conductors (Section 9.3.1) employed in the cryocable, but a different approach is required for assembling the HTS tapes. In an HTS cable the number of tapes and their pitch must be controlled to ensure nearly even current distribution among the conductor layers. The analysis in the section below addresses this problem.

Conductor Design   A generic cross section of an HTS cable is shown in Figure 9.9. It consists of the following components:

1. Former with an outer radius of R_1.
2. Layer 1 with inner and outer radii of R_1 and R_2.
3. Layer 2 with inner and outer radii of R_2 and R_3.
4. Layer 3 with inner and outer radii of R_3 and R_4.
5. Outer fictitious return conductor radius R_s used for integration.

Each layer consists of HTS tapes wrapped as shown in Figure 9.10. The tapes are wrapped with a defined pitch, and the wrapping direction is opposite between any adjacent layers.

Each conductor layer constructed with helically wrapped tapes has a self-inductance and mutual inductances with other concentric layers.

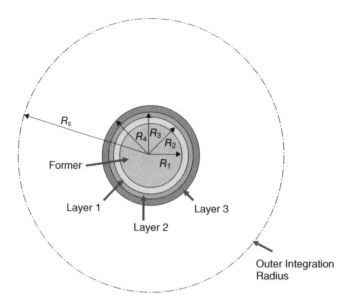

Figure 9.9   Cross section of an HTS cable

Figure 9.10   Arrangement of HTS tapes in a conductor layer

Formulas for calculation of these inductances has been published by Olsen [16]. Self-inductance (L_s) of a layer is given by

$$L_s = \frac{\mu_0}{4\pi} \tan(\theta)^2 + \frac{\mu_0}{2\pi} \ln\left(\frac{R_s}{R_2}\right) +$$
$$\frac{\mu_0}{2\pi\left(R_2^2 - R_1^2\right)^2}\left[\frac{R_2^4 - R_1^4}{4} + R_1^4 \ln\left(\frac{R_2}{R_1}\right) - R_1^2\left(R_2^2 - R_1^2\right)\right], \qquad (9.4)$$

where

L_s = self-inductance of the layer (H/m),

μ_0 = permeability of free space = $4\pi \times 10^{-7}$ (T-m/A),

R_1, R_2 = inside and outside radii of the layer (m),

R_s = radius of an outer fictitious return conductor for integration (m),

θ = twist pitch angle (radians).

In the equation above the first term represents the inductance due to the helically wrapped tapes around the former, and it is equivalent to a cylindrically wound coil. The second term represents inductance due to the linear current along the length of the conductor. The third term represents the inductance due to the magnetic energy stored within the thickness of the conductor layer. This term can be neglected (without causing significant errors in the results) because the conductor layer thickness is usually very small compared to the layer radius. In a practical conductor, the adjacent conductor layers may be separated with an insulation layer, and therefore each layer will have a unique inner and outer radii.

Likewise the mutual inductance (M_{12}) between any two layers (1 and 2) is given by

$$M = \frac{\mu_0}{4\pi}\tan(\theta_1)\tan(\theta_2)\frac{\min(R_1,R_2)^2}{R_1 R_2} + \frac{\mu_0}{2\pi}\ln\left(\frac{R_s}{\max(R_1,R_2)}\right), \qquad (9.5)$$

where

M = mutual inductance between layers 1 and 2 (H/m),

R_1, R_2 = mean radii of inner and outer layers, respectively,

θ_1, θ_2 = twist pitch of inner and outer layers, respectively.

In the equation above the first term represents coupling due to the magnetic flux linkages along the length of the cable. The second term represents coupling due to the circumferential magnetic flux.

A practical conductor may have the former built out of a stranded copper conductor (Figure 9.3) to provide for a shunt current path during a fault. The conductor may also have an outer shield layer consisting of HTS tapes wrapped with copper tapes as shown in Figure 9.3. It is also possible to include cryostat walls in the same manner in the analysis. Self-inductance and mutual inductances can be calculated for all layers using the equation (9.4) and equation (9.5).

Since all layers of a cable are shorted together at the ends, it is necessary to specify resistance for each layer for the proper distribution of transport current among all the layers. Resistance of a non-HTS tape can be calculated using its resistivity at the operating temperature of the cable with

$$R_{\text{non-hts}} = \frac{\rho_{cu}}{\pi\left(R_i^2 - R_o^2\right)} \frac{2\pi R_m}{\sin(\alpha)}, \qquad (9.6)$$

where

$R_{\text{non-hts}}$ = resistance of a non-HTS element (Ω/m),
α = twist pitch of tapes,
R_i, R_o = inner and outer radii of a layer,
R_m = mean radius of the layer.
ρ_{cu} = resistivity of copper ($\Omega \cdot$ m).

However, the resistance calculations for the HTS layers are more complex. The resistance of HTS tapes is a function of the ratio of operating to critical currents at the local field and temperature. Since the current sharing among layers may not be uniform, it is necessary to estimate the resistance of each layer as a function of local field, current, and temperature. The procedure for calculating the resistance of HTS tapes is similar to that used in Chapter 8 for fault current limiters. The specific steps for calculation of layer resistances are described below.

Since each layer is made of many HTS tapes wrapped around a mandrel, each tape experiences two field components: (1) the field along the axial length of the cable and (2) the field directed circumferentially around the conductor. The axial (B_a) and circumferential (B_c) field components are given by equation (9.7) and equation (9.8), respectively:

$$B_a = \frac{\mu_0}{2\pi R / \tan(\alpha)} I, \qquad (9.7)$$

$$B_c = \frac{\mu_0}{2\pi R} I, \qquad (9.8)$$

where

I = total current in the layer,

R = mean radius of the layer.

The field components are calculated using the current in each layer. These components—due to the various layers—can be superimposed for obtaining net the field experienced by a specific layer.

The resistance (R_{HTS}) of an HTS tape is given by equation (9.9) using the definition of a critical current, I_c (= current carried by the tape that produces a voltage drop of $1\,\mu\text{V/cm}$ in a self-field at $77\,\text{K}$):

$$R_{\text{HTS}} = 10^{-6}\,\frac{\text{V}}{\text{cm}}\left(\frac{I_0}{I_c(\theta, B)}\right)^N \frac{1}{I_0(A)}\frac{2\pi R}{\sin(\alpha)}, \qquad (9.9)$$

where

I_c = critical current at temperature (θ) and in field (B),

I_o = operating current,

N = exponent,

R = radius of layer,

α = twist pitch angle of HTS tapes

The resistance of 1-m linear length of cable can be obtained from

$$R_{\text{hts}} = \frac{1}{N_c\left[(1/R_{br}) + (1/R_{ag}) + (1/R_{sol}) + (1/R_{NiW}) + (1/R_{HTS})\right]} \qquad (9.10)$$

where

R_{br} = resistance of brass or copper stabilizer,

R_{ag} = resistance of silver layer,

R_{sol} = resistance of solder (if used),

R_{NiW} = resistance of nickel-tungsten (or hastelloy),

R_{HTS} = resistance of HTS tape,

N_c = number of tapes per layer.

Next a circuit model is constructed that includes all resistances and self-inductances and mutual inductances. An impedance matrix for a coaxial cable with a two-layer HTS HV conductor and a two-layer HTS shield is shown in equation (9.11). This matrix also includes a former (or core conductor) at the cable axis (element 1–1) and a non-HTS (copper) shield located outside the HTS shield layer 2 (element 6-6). More HTS and non-HTS layers could be included, if necessary.

$Z =$

$$
\begin{pmatrix}
R_c + j\omega L_c & j\omega M_{12} & j\omega M_{13} & j\omega M_{14} & j\omega M_{15} & j\omega M_{16} \\
j\omega M_{12} & R_{h1} + j\omega L_{h1} & j\omega M_{23} & j\omega M_{24} & j\omega M_{25} & j\omega M_{26} \\
j\omega M_{13} & j\omega M_{23} & R_{h2} + j\omega L_{h2} & j\omega M_{34} & j\omega M_{35} & j\omega M_{36} \\
j\omega M_{14} & j\omega M_{24} & j\omega M_{34} & R_{s1} + j\omega L_{s1} & j\omega M_{45} & j\omega M_{46} \\
j\omega M_{15} & j\omega M_{25} & j\omega M_{35} & j\omega M_{45} & R_{s2} + j\omega L_{s2} & j\omega M_{56} \\
j\omega M_{16} & j\omega M_{26} & j\omega M_{36} & j\omega M_{46} & j\omega M_{56} & R_S + j\omega L_S
\end{pmatrix},
$$

$$(9.11)$$

where

R = resistance of a given layer,

L = self-inductance of a layer,

M = mutual inductance with other layers,

ω = system frequency = $2\pi f$.

A common AC voltage (V_0) is applied across a 1-m length of conductor as shown in Figure 9.11. The current in each layer is given by

$$I = Z^{-1}V_0. \qquad (9.12)$$

The resistive loss in each circuit can be calculated using the individual layer's current and resistance.

HTS Coaxial Cable—High Voltage   A high-voltage single-phase coaxial cable is shown in Figure 9.3. This cable employs a copper-stranded cable as the former, a HV conductor made of two layers of HTS tapes, a shield made of two layers of HTS tapes, and a copper shield outside of the HTS shielding layer. Current sharing among various layers can be calculated with the analysis described above

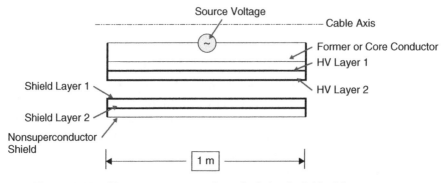

Figure 9.11   Circuit arrangement for calculating individual layer currents

(Section 9.3.2). This analysis is generic, and different numbers of layers for the HV and the shield could be used to suit the needs of a given design. The effect of components outside the cable envelope (i.e., cryostat components) could be included as separate layers. Because R_{hts} in equation (9.10) is the effective resistance of a layer (i.e., a function of current, field, and temperature), the analysis above can be used for calculating the current distribution at any transport current level, including fault currents.

An example design for a 138-kV, 2500A cable is included in Table 9.1. The HTS conductor employed has brass stabilizer soldered to the 2G tapes.

The current distribution in various layers is given below in a tabular form. The first column indicates the layer number, counting from the cable axis. The label for each layer is listed below:

0 = cable former
1 = HV HTS layer 1
2 = HV HTS layer 2
3 = HTS shield layer 1
4 = HTS shield layer 2
5 = copper shield
6 = cryostat

$n =$	LAYER CURRENTS (PU) $I_n/I_o =$	PHASE ANGLE OF CURRENTS $ang(I_n) =$
0	0.072	151.372 deg
1	0.427	−84.211
2	0.61	−88.052
3	0.434	92.331
4	0.573	91.189
5	0.024	−8.491
6	6.237×10^{-3}	−24.277

where I_n = current in the nth layer.

The second column lists the per-unit current (a fraction of the transport current) in each layer. The third column lists the phase angle of layer currents. The currents in the two-layers of the HV and the shield are approximately balanced. The number of tapes and their twist angle can be varied for obtaining a more balanced current distribution among the layers.

HTS Coaxial Cable—Medium Voltage A medium voltage, three-phase cable in a single cryostat is shown in Figure 9.12. This cable employs

Table 9.1   Coax high-voltage cable design example

Version	Coax-HV
Line voltage, kV-rms	138
Normal current, kA-rms	2.5

Cable Cross-sectional Details

Inside radius of former, mm	0.025
Outside radius of former, mm	11.1
Nominal thickness of HTS layers, mm	1
Thickness of electric insulation, mm	14.7
Outside radius of HV HTS layer, mm	12.1
Inside radius of HTS shield, mm	26.8
Outside radius of HTS shield, mm	27.2
Outside radius of copper shield, mm	29.2
LN_2 space between shield and cryostat wall, mm	5.1
Inside radius of cryostat, mm	34.3
Cryostat wall thickness, mm	24.4

Layer Characteristics

TAPE TWIST PITCH ANGLES

Copper core, deg	2.5
HV HTS layer 1, deg	−10
HV HTS layer 2, deg	15
Shield HTS layer 1, deg	−20
Shield HTS layer 2, deg	18
Copper shield, deg	−2.5

NUMBER OF HTS TAPES IN A LAYER

HV HTS layer 1	12
HV HTS layer 2	12
Shield HTS layer 1	12
Shield HTS layer 2	12

HTS Wire Details

Wire critical current at 77 K (A/cm)	300
Wire (tape) width, mm	5
HTS layer thickness, μm	1
Substrate thickness, μm	75
Silver thickness, μm	2.5
Solder thickness, μm	24.8
Brass stabilizer thickness, μm	100
Operating temperature, K	74

a copper-stranded conductor as the former, a HV conductor made of three layers of HTS tapes, a shield made of two layers of HTS tapes, and a copper shield outside of the HTS shielding layer. Current sharing among various layers can be calculated using the analysis described above (Section 9.3.2). This cable was manufactured by Sumitomo Electric using SuperPower, Inc. tapes, and it was put in service in mid-2006.

An example design for a 34.5-kV, 800A cable is included in Table 9.2. The current distribution in the various layers is given below in a tabular form. The first column indicates the layer number, counting from the cable axis. The label for each layer is listed as follows:

0 = cable former
1 = HV HTS layer 1
2 = HV HTS layer 2
3 = HV HTS layer 3

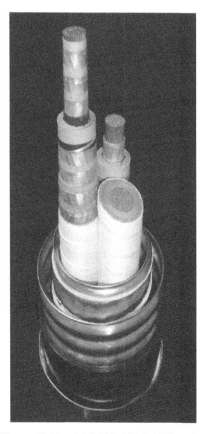

Figure 9.12   34.5-kV, 600 A HTS cable

4 = HTS shield layer 1
5 = HTS shield layer 2
6 = copper shield

The second column lists the per-unit current (a fraction of the transport current) in each layer. The third column lists the phase angle of layer currents. The currents in the three layers of the HV and the shield are approximately balanced. The number of tapes and their twist angle can be varied for obtaining a more balanced current distribution among the layers.

$n =$	$I_n/I_o =$	$\text{ang}(I_n) =$	
0	0.075	154.471	deg
1	0.195	−77.826	
2	0.335	−89.834	
3	0.506	−86.757	
4	0.471	92.485	
5	0.566	92.023	
6	0.053	−42.775	

Triax™ HTS Cable—Medium Voltage   A low-voltage three-phase Triax™ cable is shown in Figure 9.4. Such a cable has been in operation since 2007, and a few more are under construction. This cable employs a hollow former, individual phase HV conductors made of a layer of HTS tapes, and a copper shield outside of the last HTS layer. Current-sharing among various layers can be calculated using the analysis described above (Section 9.3.2). Parameters (resistance, self-inductance and mutual inductances) are calculated with the formulas used for the other example cases. The applied voltages to the three phases are phase separated by 120° for a balanced three phase system.

The current distribution in various phases (layers) for a 13.8-kV, 1250A cable is given below in a tabular form. The first column indicates layer number counting from the cable axis. Label for each layer is listed as follows.

0 = cable former
1 = HV HTS Ph-A layer
2 = HV HTS Ph-B layer
3 = HV HTS Ph-C layer
4 = copper shield layer

Table 9.2   Coax medium voltage cable design example

Version	Coax-MV
Line voltage, kV-rms	34.5
Normal current, kA-rms	0.80

Cable Cross-sectional Details

Inside radius of former, mm	1
Outside radius of former, mm	8
Nominal thickness of HTS layers, mm	1.5
Thickness of electric insulation, mm	4.4
Outside radius of HV HTS layer, mm	9.5
Inside radius of HTS shield, mm	13.9
Outside radius of HTS shield, mm	15.4
Outside radius of copper shield, mm	17.4

Layer Characteristics

TAPE TWIST PITCH ANGLES

Copper core, deg	2
HV HTS layer 1, deg	28
HV HTS layer 2, deg	10
HV HTS layer 3, deg	−24
Shield HTS layer 1, deg	25
Shield HTS layer 2, deg	−20
Copper shield, deg	2

NUMBER OF HTS TAPES IN A LAYER

HV HTS layer 1	10
HV HTS layer 2	10
HV HTS layer 3	10
Shield HTS layer 1	18
Shield HTS layer 2	18

HTS Wire Details

Wire critical current at 77 K (A/cm)	250
Wire (tape) width, mm	5
Substrate thickness, μm	75
Silver thickness, μm	2.5
Operating temperature, K	73

The second column lists per-unit current (fraction of transport current) in each layer. The third column lists the phase angle of layer currents. The currents in the three phases are balanced. The tape twist angles are varied for obtaining a balanced current distribution among layers.

Table 9.3   Triax medium voltage cable design example—Single layer

Version	Triax-MV
Line voltage, kV-rms	14
Normal current, kA-rms	1.26

Cable Cross-sectional Details

Inside radius of former, mm	10
Outside radius of former, mm	11
Nominal thickness of HTS layers, mm	1
Thickness of electric insulation, mm	2.2
Outside radius of HV HTS layer, phase-A, mm	14.2
Outside radius of HV HTS layer, phase-B, mm	17.4
Outside radius of HV HTS layer, phase-C, mm	20.7
Outside radius of copper shield, mm	22.9
LN_2 space between shield and cryostat wall, mm	5
Inside radius of cryostat, mm	27.9
Cryostat wall thickness, mm	25

Layer Characteristics

TAPE TWIST PITCH ANGLES

Copper core, deg	2
HV HTS layer phase-A, deg	−5.0
HV HTS layer phase-B, deg	26.3
HV HTS layer phase-C, deg	−10.0
Copper shield, deg	−2.5

NUMBER OF HTS TAPES IN A LAYER

HV HTS layer phase-A	16
HV HTS layer phase-B	16
HV HTS layer phase-C	16

HTS Wire Details

Wire critical current at 77 K (A/cm)	300
Wire (tape) width, mm	5
HTS layer thickness, μm	1
Substrate thickness, μm	75
Silver thickness, μm	2.5
Solder thickness, μm	24.8
Brass stabilizer thickness, μm	100
Operating temperature, K	72

LAYER CURRENTS (PU)		PHASE ANGLE OF CURRENTS	
$n =$	$I_n/I_o =$	$\mathrm{ang}(I_n) =$	
0	2.072×10^{-4}	61.931	deg
1	1.002	−88.626	
2	1.003	148.279	
3	0.994	30.609	
4	5.243×10^{-3}	−50.216	

The number of tapes and their twist angle can also be varied for obtaining a more balanced current distribution among the layers. More tapes in a phase and a larger former diameter are required to carry the larger current.

9.4 CHALLENGES

Underground transmission cables (69 kV and higher) with cross-linked polyethylene (XLPE) insulation have been in use worldwide since the 1960s. Today, underground XLPE transmission technology is established worldwide up to 400 kV, with major projects in Copenhagen, Berlin, and other metropolitan areas and up to 500 kV in Japan. Underground transmission lines can be built where there is no space for overhead lines, thus reinforcing the transmission system where necessary. They can cross cities and bodies of water. They are extremely reliable and not vulnerable to ice or wind storms, sagging due to overload, tree growth, and similar environmental factors. In the United States the power cables are designed, manufactured, and installed using XLPE cable technology up to 345 KV. The biggest challenges for both cryocable and HTS cable systems are to compete with these highly reliable, maintenance free XLPE cable systems.

From user's perspective, some of the important considerations for product selection are the lowest cost without diminishing the level of reliability, the lowest cost solution (including capital and life-cycle costs) and the best value for customers, recovery of investments in a timely manner, and reliability of the cryogenic system. These challenges are addressed below.

9.4.1 Resistive Cryogenic Cable

Although a cryocable employs a conventional aluminum conductor to carry current, the I^2R losses in the conductor are significant. These losses must be removed with a refrigerator with a penalty of over 1:8

(smaller capacity G-M coolers have the penalty ratio of 1:30). The overall operating cost of a cryogenic cable system is similar to that of a conventional cable system of equal rating. For example, a 500-kV, 1200A cable could be built and installed using conventional cable installation techniques. A large dedicated LN$_2$ refrigerator could cool a 30-mile section of the cryogenic cable. However, three conventional force (oil) cooled cables are required to carry the same load, which requires a more than three times wider right of way. The capital and operating costs of cryogenic cable and three-conventional cables are similar. The only benefit of cryogenic cable appears to be the one-third the right-of-way requirement, which could be an important consideration for many inner city applications. Another significant benefit relates to large fault current-carrying capability, which is difficult to achieve in HTS cables. A cryogenic cable is a more reliable solution if an operating cost similar to that of a conventional cable of similar rating is acceptable. This cable, constructed like a conventional cable (with both conductor and insulation) and operating at the LN$_2$ temperature could have a much longer life than a conventional oil-filled cable, and therefore its lifetime cost could be significantly lower than that of a conventional cable system. To the author's knowledge, this concept is not being developed by any organization at the present time.

Concern about Large Concentration of LN$_2$ Nitrogen is a colorless, odorless, stable, and nonflammable gas that occurs naturally in the atmosphere. It constitutes approximately 79% of the earth's atmosphere and, under normal conditions, poses no threat to health or safety. When cooled, nitrogen condenses into a liquid at a boiling point of $-195.8°C$ ($-320.4°F$, or $77K$); it is abundant in nature and routinely made available as a by-product of oxygen liquefaction (the boiling point of oxygen is $90K$). Because it is abundant, low-cost, noncorrosive, nonflammable, and has high electrical insulating properties, liquid nitrogen is an ideal coolant and dielectric for electric power applications of cryogenic and HTS cables.

There are two principal risks associated with nitrogen: (1) oxygen deficiency and (2) extreme low temperature. The release of high concentrations of nitrogen into the environment, in confined spaces, can result in asphyxiation by displacement of oxygen. Raised concentrations of nitrogen may cause a variety of respiratory symptoms, and at high concentrations, unconsciousness or death may occur. Direct contact with liquid nitrogen can cause frostbite-type injuries, and must be avoided. The low temperature of liquid nitrogen can also result in a temporary fog that will lower temperatures and obscure vision until it dissipates. If a spill or leak occurs in a nonconfined space, nitrogen will rapidly dissipate into the environment. There is no known adverse health effects associated with chronic exposure to this gas.

In order for cryocable and HTS cable to be adopted on a widespread basis, several design options are available for minimizing the risk of accidental exposure to liquid nitrogen or excessive concentrations of gaseous nitrogen. These include the following:

- The area above an HTS cable should be protected with some means of deflecting backhoe hits, in much the same way that conventional underground cables and high-pressure gas pipelines are protected and marked. Concrete, wood, or steel barriers, with an audible and/or visual warning system in case of physical contact, would reduce the incidence of accidental dig-ins.
- A system of sensors along the length of a LN_2-filled cable system could be employed to detect low temperatures, low concentrations of oxygen, or both.
- Careful monitoring of coolant flow, temperature, and pressure can be used to determine the presence of a leak or fault.

Sensor systems could be employed to detect slow leaks, drops in vacuum pressure, or catastrophic failures along the entire length of an HTS cable, using low-power remote telemetry capable of being powered by solar panels. In the event of a system failure, a signal can be sent to issue an alarm and/or shut off the flow of liquid nitrogen into the cable if this is deemed appropriate.

9.4.2 HTS Cable

The most attractive application of HTS cables is in inner cities. For widespread adaption by utilities to occur, they must satisfy or address the following five criteria:

1. Low capital and operating cost.
2. Refrigeration and cooling system reliability.
3. Performance during normal and fault periods.
4. Availability.
5. Fault current limiting cable (option).

Capital and Operating Cost HTS wire, cryostat, and refrigerator costs are currently high. The total capital system cost of an HTS system could be three to four times that of a conventional system of similar rating. Despite the compactness of such cables the capital cost could be a major deterrent to their applications in an electric grid. Manufacturers have set a cost goal of <$10/kA-m for the second generator (2G) since 1995, but it has not been achieved yet. For commercial success, the HTS wire price must drop to $5 to $10/kA-m range.

In short-length inner city applications, the heat leak through the cable cryostat and the terminations losses present the largest thermal loads. The cryogenic refrigerator must operate at nearly its rated capacity regardless of the cable load current. This adds to the cable system's operating cost.

Refrigeration and Cooling System Reliability   The cooling system for an HTS cable is highly critical. In the event of a cooling system failure, an HTS cable could carry the nominal load only for a short time. Thus the cooling system for HTS cables must be highly reliable at affordable cost. Currently the available refrigerator systems are costly, and their long-term reliable operation is less than desirable. For smaller systems the Gifford-McMahon type of coolers could be employed, but their current cost ($30,000 for 300-W cooling at 77 K) is high. Several coolers could be combined to achieve the required cooling power. The reliability issue could be addressed by an N-1 redundancy of coolers, but they must be serviced manually when they fail. The cooling system is the most critical component of the HTS cable system because the cable is unable to carry any significant load unless it operates close to the nominal operating temperature.

Thus the necessity to maintain the cooling at the cryogenic (liquid nitrogen) temperatures presents a failure mode for the HTS cable system. This fact requires a highly reliable design for HTS cable cooling systems, both at the component level and at the system level. At the component level, conventional coolers to achieve liquid nitrogen temperatures have been employed widely for several decades, with a high degree of reliability. Cryocooler technology is employed across a number of industries. To facilitate broader use of cryocooler technology, work is underway to develop lower cost, more reliable, modular pulse-tube refrigeration systems with the fewest possible moving parts.

Performance during Normal and Fault Periods   Because of the high cost of HTS wire, its content in a cable is usually minimized. The amount of HTS employed is usually sufficient to carry an overload current (up to about 2×) specified by a customer. However, a cable system could experience currents levels of 40 to 80 kA. Such high current faults cause the HTS wires to quench. Once quenched, the joule heating raises the temperature of the wire further and forces a shut-down of the cable system. A cable in a utility grid may experience many faults in the range of 30 to 40 kA. A cable system is expected to keep operating during these events. However, an HTS cable designed for 2× overload may get partially quenched if it experiences even a 35-kA fault, for example. Many such low-fault current events are the result of starting large motor loads (like railroad subway cars in inner cities) and transformer switching operations. Also the cable may not quench uni-formly along the length of the cable.[§] A hot spot created by local quench could lead to a cable burnout. A long-term operating experi-ence is needed with these cables to establish their design and opera-tional rules precisely.

Another disadvantage of HTS cables is that they cannot be imme-diately restored to service after experiencing a high current fault or a repair to a circuit because of the need to reach cryogenic temperatures to achieve the superconducting state. Higher voltage cables are more difficult to cool due to the thick covering of the HTS conductor by the dielectric, which also has a very poor thermal conduction property. HTS cables therefore require contingency backups for such situations. While the HTS cable is unable to serve its load, service to customers in the vicinity of an HTS cable must be provided through alternate pathways until the design operating temperature is achieved.

Availability   All HTS cables take a long time to cool down from room temperature to their operating temperature. This cool-down period could be in the range of several days to several weeks. After experienc-ing a high current fault, these cables can also warm up sufficiently to force a shutdown for several hours to days. During this period alternate routes must be available to carry the load that was previously carried by the HTS cable. This has a major impact on the availability of HTS cables to the system operators that serve loads during critical periods. The utility operators have to provide an alternate route to carry the power. Thus, until their reliability is established, an HTS cable can only play a supportive role in an electric grid.

The HTS cables have also been proposed for tying offshore wind farms with HTS cables. This is a bad idea. Wind power output is

§ Unlike low-temperature superconductors, a quench of HTS conductor does not prop-agate well along the length of a conductor.

unreliable, and HTS cable reliability is also questionable due to the refrigeration and fault current issues that could arise. Two unreliable systems in series will have a huge adverse impact on the reliability and availability of the overall system.

FCL Cable   Some HTS cable proponents are proposing to incorporate fault current limiting capability in the HTS cable. One such cable is being prototyped for application in the Con Ed grid in New York City. However, such cables are likely to further reduce their availability during critical periods. The FCL cables employ a flux flow resistivity region of the HTS wire when the wire resistance goes up. This increased resistivity could be used for limiting the fault current. However, a minimum length (based on the voltage experienced by the cable during a fault and the current limiting requirement) is needed to achieve the desired fault current limiting action. Moreover the uniform transition of HTS wire to the flux flow region that is required cannot be guaranteed. Thus it is not clear how well such a cable would perform in a grid. A cable may develop local hot spots where it could heat uncontrollably and become a fuse. Performance of such a system must also be critically studied in a grid before adapting it as the primary mode of a power carrier.

Coaxial HTS cables with HTS shields are also characterized as very low impedance (VLI) cables [1]. This raises the question of whether the low impedance characteristic of VLI superconductor cables is maintained during high current faults in a grid. For example, during a fault the cable current is expected to reach up to 60 to 80 kA. It is a fundamental property of HTS conductors that under fault conditions they immediately transition to a normal state once the current exceeds the critical current. As the current increases, this normal state is initially a moderately low resistivity state called the flux flow state, but once the current exceeds several times its rated current, the full normal state resistivity of order $100 \mu\Omega$-cm is reached. Under fault conditions these transformations can occur within milliseconds; that is well within a single AC cycle. Then all current is shifted to the copper core (and copper shield wires, if necessary). Thus, during a low current fault which does not transition HTS to its normal state, the VLI characteristic of the cable is maintained. However, during high current faults, the HTS cable effectively becomes a normal cable with cryogenic copper and has higher impedance similar to that of a conventional cable. Also, as stated above, once a cable quenches, a long time is needed to cool it back to its normal operating temperature. This requires other means to supply the load previously served by the HTS cable.

9.5 MANUFACTURING ISSUES

Many HTS cable have been prototypes and tested successfully. Most cables were manufactured using conventional cable machines with minor modifications. Cables with their cryostats, splicing, and termination techniques have been successfully demonstrated. Even variations of ground elevations have been address for inner city applications. This section discusses issues relating to manufacturing of cryocable and HTS cables.

9.5.1 Resistive Cryogenic Cable

The cryocable with its aluminum conductor is easiest to manufacture as it can use existing cabling machinery. However, it also requires a vacuum-insulated pipe to house the cable. This pipe has a cost impact, but manufacturability is not an issue. The cable termination and splicing techniques are similar to those of conventional cables, with some minor modifications due to the involvement of LN_2 cooling.

A disadvantage of both resistive cryocable and HTS cables is that they cannot be immediately restored to service after a repair to a circuit because of the need to reach the required cryogenic temperatures. Cryogenic coolant reserves and access to the extensive industrial gas infrastructure for the supply of cryogenic liquids can reduce the cooldown period.

The necessity to maintain cooling to cryogenic (liquid nitrogen) temperatures introduces a new type of failure mode for the HTS cable. This fact requires a highly reliable design for HTS cable cooling systems, both at the component level and at the system level. At the component level, conventional coolers to achieve liquid nitrogen temperatures have been employed by thousands of users for several decades, with a high degree of reliability. Many of these coolers are in unattended operation at remote locations, in conditions similar to those that would typify utility grid operations. Cryocooler technology is employed across a number of industries, and widespread utility adoption of HTS cable (e.g., several thousand miles of cable) would result in an incremental increase in the user base. To facilitate broader use of cryocooler technology, work is underway to develop lower cost, highly reliable, modular pulse-tube refrigeration systems with the fewest possible moving parts. At the system level, cooling networks will likely need to be designed that incorporate redundancy, reflecting the same N-1 type of approach that is reflected in overall transmission system design. Existing business models, for example, involving the outsourcing of reliable cooling systems, supported by existing infrastructure and managed by industrial gas suppliers, can be used to ensure that HTS cable systems are operated reliably and at least cost. All in all, cooling technology is sufficiently well developed and widely deployed in industry that this is not seen as a major obstacle to achieving a highly reliable transmission cable; however, appropriate systems must be designed and tested at the fully commercial level.

9.5.2 HTS Cable

HTS cable manufacturing issues relate to fabrication of the HV HTS conductor, HTS ground shield, wrapping of HV conductor with dielectric insulation, continuous cryostat fabrication, splicing, terminations, and cooling system. The cooling system issue has been discussed in the previous section in relation with the resistive cryogenic cable. Other issues are discussed below.

HTS Conductor All HTS conductor configurations and their manufacturing issues have been discussed in Chapter 2. Both BSCCO-2223 and the YBCO-coated HTS conductor are suitable for cable manufacturing. HTS conductors sourced from various manufacturers have been successfully employed for constructing cable prototypes worldwide.

Dielectric Applications The most popular material for insulating these cables has been Kapton and Nomex (DuPont Products), or the equivalent polypropylene-cellulose paper laminate (PPPL). These materials are applied by lapping a necessary number of about 1-inch-wide tapes using techniques commonly employed for conventional cables.

Continuous Cryostat The flexible vacuum-insulated cryostat shown in Figure 9.13 consists of two flexible and concentric corrugated tubes made of austenitic stainless steel with a reduced carbon content. The outer part of the inner tube is covered with layers of super-insulation. A spacer with low thermal loss centers the inner tube inside the outer tube and prevents contact between the two metallic tubes. A molecular sieve guarantees a long-term vacuum. The cryostat is protected with a polyethylene jacket. The cryostats, together with their associated terminations and hardware, are assembled, leak tested, and evacuated at the factory. This permits a simple and cost-saving site installation. The

Figure 9.13 Continuous cryostat thermal insulation for HTS conductors

flexibility of the cryostat permits handling like conventional cables. Flexible cryostats have been successfully built and installed in a number of cable prototypes around the world.

Terminations Each cable length has two terminations to connect the cold end of the cable to the bushing interface with the room-temperature components. Inside the terminations a current lead connects the superconducting cable conductor to a normal conducting bushing. The current lead handles the transition to room temperature by taking into account the dielectric requirements of the cable and of the terminations while minimizing thermal conduction from room-temperature end to the cold end. The terminations have to handle the flow of liquid nitrogen going through the cable and back. These terminations have been successfully built and installed for prototype HTS cable for voltages up to 138 kV.

Splicing Cable splices are needed because only a limited length of cable can be shipped from the factory. A splice must accommodate the superconductor to superconductor joint with minimal resistance and must provide for through and fro movement of the LN_2 coolant. Some of the splices would have to accommodate injection and extraction of the LN_2 coolant to interface with an intermediate cooling station. Such joints could be challenging for Triax™ type cables where the LN_2 flows through the bore of the cable. Development of these joints is being carried out as part of ongoing development projects.

Field Repair A cable can experience damage for a variety of reasons such as earth digging in cable vicinity or damage of the cable due to electrical reasons. It would be necessary to develop procedures and tools to carry out such repair work in the field. It is not clear how much attention has been devoted to this aspect so far.

9.6 PROTOTYPES

Many HTS power cable projects have been prototyped successfully around the world. Three major projects in the United States are the 13-kV Triax™ cable in Columbus, Ohio; the 34.5-kV, three cables in a common cryostat in Albany, New York; and the 138-kV, coaxial cable on Long Island, New York. In addition to these prototypes, two current projects are a Triax™ cable installation in New Orleans and an HTS Triax™ cable with FCL capability in Con Ed system in New York City. Besides power-transfer capability, the Con Ed cable includes the fault-limiting feature of HTS cable technology [17]. A high-resistance stabilizer is employed in the HTS wire. Under normal conditions the current is carried in the HTS layer, but when a fault occurs, the high-resistance layer comes into play, reducing the fault current. Once the fault has passed, the normal load current transfers to the HTS layer again. This cable is utilized for power transfer between distribution substations. Top level description of these prototypes is provided below.

9.6.1 Resistive Cryogenic Cable

The resistive cryogenic cable was prototyped and demonstrated in the late 1970s. The cable was rated 500 kV, 1200A. A prototype termination built by GE for the cable is shown in Figure 9.14. The cable employed an aluminum conductor cooled with LN_2. A single cable could carry 3× more power than a conventional oil-cooled cable of similar size (outside diameter). Subsequently a joint study was conducted with PSE&G of New Jersey for a grid application. The technical and economical feasibility was confirmed. The present-day capital costs of three conventional cables versus a single cryogenic cable were determined to be essentially identical. However, PSE&G felt that it was risky to allocate a large chunk of power to a single link because, in the event of a sudden interruption of this link, the whole electric grid might get unstable and initiate a cascading blackout. This concern is still valid for electric utilities for both cryocable and HTS cables.

9.6.2 HTS Cable—High Voltage

American Superconductor and Nexans have built and installed a 138-kV, 2500A HTS cable in the Long Island Power Authority (LIPA) grid

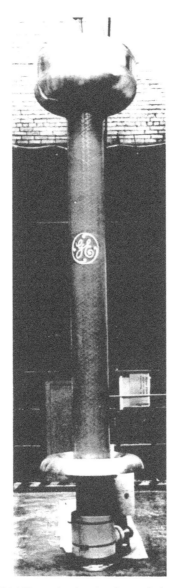

Figure 9.14 500-kV termination for the resistive cryogenic cable

on Long Island, New York. This cable employs 1-G HTS conductor in a single phase coaxial configuration. Figure 9.15 shows the cable installation. The purpose of this project [6] is the demonstration of a high temperature superconducting (HTS) power cable in the Long Island Power grid, spanning nearly half a mile and serving as a permanent link in the Long Island Power Authority's (LIPA) grid network. The cable represents the world's first installation of an HTS cable in a live grid

Figure 9.15 AMSC/Nexans's high-voltage cable installation on Long Island, New York

at transmission voltages. The cable has been installed in an existing right-of-way in Lake Ronkonkoma, New York. The 138-kV cable is capable of delivering power to 300,000 homes and is an integral part of the LIPA grid. The cable was energized on April 22, 2008. LIPA plans to retain this HTS cable as a permanent part of its grid.

9.6.3 HTS Cable—Medium Voltage

SuperPower, Inc. and Sumitomo built and installed this 34.5-kV, 800A cable in National Grid in Albany, New York. This cable employs 1G HTS conductor over most of its length and has a 2G HTS cable section incorporated at a later date. Figure 9.16 shows the cable installation. The purpose of this cable project [7] is to demonstrate an HTS cable in the power grid of Albany, New York, that includes a first-of-a-kind application of a cable splice and a section of cable fabricated using 2G superconducting wire.

Figure 9.16   Medium-voltage SuperPower/Sumitomo's cable installation in Albany, New York

The cable employs "three-core HTS cable technology," meaning that the cable had three separate cores (copper) in a single cryogenic pipe (the *cryostat*). Each core is surrounded by layers of HTS wire and electrical insulation, and the whole assembly is surrounded by liquid nitrogen coolant and thermal insulation. The design also uses a "cold dielectric" scheme, in which the cryogenic fluid and thermal insulation surround the electrical insulation in the cable. The cable is designed to carry 800 amps at 34.5 kV. The cable is 350 meters long and is installed underground in Albany, New York, between two National Grid substations near the Hudson River and under Interstate 90. The project demonstrated the first-ever splice in an HTS cable. In the second phase of the project, a 30-meter section was replaced with a section of cable made with 2G HTS wire manufactured by SuperPower. The new wires have both performance and cost advantages over earlier HTS wires and are expected to hasten HTS technology's entry into the cable market. In parallel, a manufacturer of cryogenic systems (Linde) developed a refrigeration system that meets the stringent reliability and efficiency standards required by the utility industry.

The first phase of the project was energized on July 20, 2006. It operated flawlessly as an integral part of National Grid's 35-kV network in Albany, serving the equivalent of 25,000 homes. It was taken offline

Figure 9.17   Low-voltage ULTERA cable installation in Columbus, Ohio

after nine months to begin phase 2 of the project: replacing a 30-meter section with 2G cable. Phase 2 of the project was energized on January 8, 2008, and was taken offline after approximately 2400 hr of successful operation. This was the world's first installation of a device using 2G wire in a utility grid.

9.6.4   Triax™ HTS Cable—Medium Voltage

ULTERA (a partnership between US cable manufacturer Southwire and nkt cables of Denmark) built and installed this 13.2-kV, 3000A cable in American Electric Power (AEP) grid in Columbus, Ohio. This cable employs 1G HTS conductor. Figure 9.17 shows the cable installation. The purpose of this project [8] is to field-test a long-length HTS cable under real environmental stresses and real electrical loads. The cable system forms an important electrical link in an AEP substation in Columbus, Ohio. The HTS cable, carrying three to five times more power than conventional cable, can meet increasing power demands in urban areas via retrofit applications, eliminating the need to acquire new rights-of-way.

This cable incorporates all three phases of a power line in a single cable. This reduces the cooling load of the system, and the concentric phases combine to reduce stray electromagnetic fields. In addition this design requires only about half the superconducting materials required by coaxial designs and takes up less space, since a single cable performs the job of three separate cables.

This 200-meter cable was energized August 8, 2006; it has operated successfully since, serving the equivalent of 36,000 homes. Most of the cable was pulled into conduit underground, and a cable splice was built in a man-hole to demonstrate the joining process for multiple cable sections. The cryogenic requirements are met by a base vacuum system, while a new pulse-tube cryocooler provides approximately 40% of the cooling requirement and demonstrates high reliability and low maintenance for future commercial applications.

ULTERA is also building a 13.8-kV, 2000A cable for installation in the Entergy grid in New Orleans, Los Angeles. Once energized in 2011, the HTS Triax cable located in the New Orleans area will deliver the needed power in a growing area while allowing Entergy Louisiana to avoid building a new, high-voltage step-down substation in a congested city. The HTS system will allow for a small footprint, environmentally friendly, highly efficient HTS cable system. Entergy Louisiana will install 1.1 miles of cable—the longest HTS cable system in the world to date—to connect the Labarre substation to the Metairie substation outside of New Orleans to provide power to the growing suburban area.

This project will attempt to prove that installation and operation of HTS cables in lengths greater than 1 mile are possible. Entergy Louisiana will implement a "virtual substation" concept and allow the utility to avoid construction of a costly new 230-kV substation. The project will also demonstrate the efficiency and cost savings of the distribution voltage HTS Triax cable compared to the conventional transmission voltage circuit, which it will replace.

Another ULTERA project is for the Con Ed grid in Manhattan, New York. The plan is to install a 13.8-kV, 4000A cable that will be operational by the end of 2010. This cable will incorporate the fault current limiting feature. Known as Project Hydra, the installation of 300 meters of HTS Triax cable in Con Edison's power grid will demonstrate the ability of HTS technology to relieve system congestion and reduce the costs of power delivery in densely populated urban areas. This project will utilize the highest current rating to date, at 4.0 kA while implementing a "substation bus tie" application to allow the utility to leverage transformer assets between distribution substations. This is to have a low impact to Con Edison and the Manhattan area—that is, a small equipment footprint, no costly new transformers, smaller right-of-way requirements, and little disruption of residential areas. The application in one of the most densely populated US cities areas increases the potential of HTS cables to help solve the looming electric power challenges and security risks many cities will face in the near future.

9.7 SUMMARY

All cables employing LN_2 coolant include a vacuum-insulated cryostat and high-cost refrigeration systems. The operating costs of the refrigeration and resistive losses in circuits are important components in an economic evaluation of HTS cable systems. Because of high component costs, such cables are economically feasible only as high current (power) links. Fear of losing a high-power link and initiating grid-wide blackout is still a major concern deterring the adaptation of this technology by electric utilities worldwide. The initial application of HTS technology is expected to be in innercity grids where space limitations and smaller power links constitute incentives for their adaptation. Nevertheless, the biggest challenge for cryocable and HTS cable systems is in these systems' capabilities to compete with highly reliable, maintenance free XLPE cable systems.

REFERENCES

1. J. Howe, B. Kehrli, F. Schmidt, M. Gouge, S. Isojima, and D. Lindsay, "Very Low Impedance (VLI) Superconductor Cables: Concepts, Operational Implications and Financial Benefits," Whitepaper, November 2003, American Superconductor Corp. www.amsc.com/products/library/vli_cable_white_paper_nov03.pdf.

2. M. Nassi, N. Kelley, P. Ladie, P. Corsato, G. Coletta, and D. von Dollen, "Qualification Results of a 50 m-115kV Warm Dielectric Cable System," *IEEE Trans. Appl. Superconductivity* 11 (1, Pt. 2): 2355–2358, 2001. DOI 10.1109/77.920334.

3. P. W. Fisher, M. J. Cole, J. A. Demko, C. A. Foster, M. J. Gouge, R. W. Grabovickic, J. W. Lue, J. P. Stovall, D. T. Lindsay, M. L. Roden, and J. C. Tolbert, "Design, Analysis, and Fabrication of a Tri-axial Cable System," *IEEE Trans. Appl. Superconductivity* 13 (2, Pt. 2): 1938–1941, 2003. DOI 10.1109/TASC.2003.812969.

4. S. Honjo, M. Shimodate, Y. Takahashi, T. Masuda, H. Yumura, C. Suzawa, S. Isojima, and H. Suzuki, "Electric Properties of a 66 kV 3-Core

Superconducting Power Cable System," *IEEE Trans. Appl. Superconductivity* 13 (2, Pt. 2): 1952–1955, 2003. DOI 10.1109/TASC.2003.812975.

5. S. K. Olsen, O. Tonnesen, and J. Ostergaard, "Power Applications for Superconducting Cables in Denmark," *IEEE Trans. Appl. Superconductivity* 9, (2, Pt. 1): 1285–1288, 1999. DOI 10.1109/77.783536.

6. Long Island HTS Power Cable, Factsheet May 16, 2008. US-DOE Office of Electricity Delivery and Energy Reliability.

7. Albany HTS Power Cable, Factsheet July 24, 2008. US-DOE Office of Electricity Delivery and Energy Reliability.

8. Columbus HTS Power Cable, Factsheet February 20, 2008. US-DOE Office of Electricity Delivery and Energy Reliability.

9. E. B. Forsyth, "Underground Power Transmission by Superconducting Cable," Brookhaven National Laboratory, Upton, NY, BNL 50325, March 1972.

10. E. B. Forsyth and R. A. Thomas, "Performance Summary of the Brookhaven Superconducting Power Transmission System," *Cryogenics* 26(November): 599–614, 1986.

11. S. S. Kalsi, "Cryogenic Cable—Tomorrow's Answer to 'No Room' Upstairs," *GE Forum* 1(1): 1975, pp. 27–30.

12. Resistive Cryogenic Cable Economic Viability Evaluation, Final Report prepared by GE for US-DOE, Report CCS-80-4, March 31, 1980.

13. S. S. Kalsi and M. Hudis, "Cryogenic Insulation Development for High Capacity Underground Cable," *IEEE Power Apparat. Syst.* 93: 1974, pp. 558–566.

14. S. S. Kalsi and S. H. Minnich, "Calculation of Circulating Current Losses in Cable Conductors," *IEEE Power Apparat. Syst.* 99: 558–563, 1980.

15. Resistive Cryogenic Cable, Phase II Final Report prepared by General Electric under contract 14-01-0001-1483 for Edison Electric Institute, US Department of the Interior and Tennessee Valley Authority, April 1974.

16. S. K. Olsen, C. Traeholt, A. Kuhle, O. Tonnesen, M. Daumling, and J. Ostergaard, "Loss and Inductance Investigations in a 4-Layer Superconducting Prototype Cable Conductor," *IEEE Trans. Appl. Superconductivity* 9 (2, Pt. 1), 833–836, 1999. DOI 10.1109/77.783426.

17. G. Wolf, "Technologies Advance Transmission Systems," *Transmission Distrib. World*, Penton Media, Nov. 2007, TD-175-SPH.

10

MAGLEV TRANSPORT

10.1 INTRODUCTION

Magnetically levitated (Maglev) trains have the potential to be faster, quieter, and smoother than current wheeled mass transit systems. These trains have been a popular subject of a future mode of travel for a very long time, but it appears that the time for practical implementation is near. Economics continues to be the main driver in their ready acceptance and implementation. Maglev construction projects face many complex issues—financing (public or private), right-of-way, ownership, job displacement from current modes of transportation to the future Maglev, are but a few. The transformation from horse-pulled buggies to railroad was perhaps simpler, but a change from railroad to Maglev is proving difficult because of entrenched interest of labor in the current railroad system and lack of public funding. Of course, the introduction of Maglev trains would be capital intensive, yet so was the railroad. Railroads were initially funded privately. They were only gradually taken over by public funding, as their reliability could not be left at the mercy of business profit/loss logistics. By the same reasoning, almost all major airports were publically funded. Likewise, in modern times, with uncertain business and energy markets, it is prudent to consider public

Applications of High Temperature Superconductors to Electric Power Equipment, by Swarn Singh Kalsi
Copyright © 2011 Institute of Electrical and Electronics Engineers

sources for deployment of Maglev technology. The same as with airports, public funding could be used to build guideways, and like airplanes, Maglev vehicles could be privately owned and operated.

The purpose of this chapter is to describe magnet technology employed in Maglev systems. Maglev trains differ from conventional trains in that they are levitated, guided and propelled along a guideway by a changing magnetic field rather than by steam, diesel, or electric engine. The absence of direct contact between the train and the rail allows the Maglev to reach record ground transportation speeds, comparable to that of commercial airplanes. Various Maglev train concepts have been reviewed by Lee [1] and Thornton [2]. This chapter reviews the major Maglev concepts, describes the benefits and challenges of HTS Maglev technology, and includes a short discussion of the current prototypes and demonstration projects.

10.2   CONFIGURATIONS

There are two primary types of Maglev configurations:

- *Electrodynamic suspension (EDS).* Electromagnets keep the vehicle levitated off the track by use of magnets on both track and vehicle (pushing the vehicle away from track).
- *Electromagnetic suspension (EMS).* Actively controlled electromagnets keep the vehicle attracted to a magnetic iron rail fixed to the track.

For slow-speed applications a stabilized permanent magnet suspension (SPM) employs opposing arrays of permanent magnets to levitate the vehicle above the rail. Stabilization is achieved with actively controlled trim magnets. EMS systems are also available for slow-speed applications. These slow speed systems are not discussed in this book.

Descriptions of the two main EDS and EMS concepts suitable for high-speed applications are provided below.

10.2.1   Electrodynamic Suspension (EDS)

Japan Railways (JR) has been developing the concept of electrodynamic suspension for a long time and has constructed a nearly 20-km test track. The JR-Maglev levitation train uses the EDS system shown in Figure 10.1. Moving magnetic fields create a reactive force repelling the currents induced in the conductors or coils in its vicinity. This force

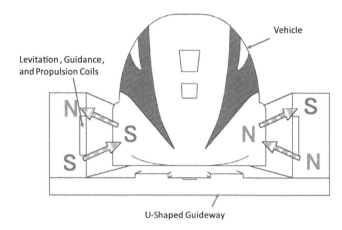

Figure 10.1 Electrodynamic suspension (EDS) Maglev concept

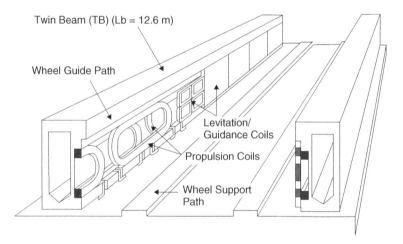

Figure 10.2 Conceptual arrangement of levitation/guidance and traction coils

holds up the train. The Maglev trains have superconducting magnetic coils, and the guideways contain passive levitation coils. When the trains run at high speed, the currents induced in levitation coils on the guide way produce reactive forces in response to the approach of the superconducting magnetic coils onboard the trains.

EDS has the advantage of larger gaps (~100 mm) but requires support wheels during low-speed running. EDS does not produce a large levitation force at low(er) speeds (150 km/h or less in JR-Maglev). However, once the train reaches a certain speed, the wheels retract and the train levitates. A conceptual arrangement of levitation/guidance and traction coils in the guideway is shown in Figure 10.2.

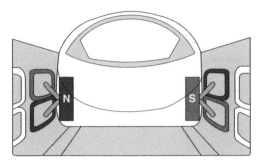

Figure 10.3 Levitation coils in sidewalls of U-shaped track

Principle of Magnetic Levitation The 8-figured levitation coils installed on the sidewalls of the guideway are shown in Figure 10.3. When the on-board superconducting magnets pass at a high speed several centimeters below the center of these coils, an electric current is induced in the levitation coils. Interaction between the levitation coils and the superconducting magnet push the vehicle upward. During stable levitated operation, the upward pulling force of the upper figure-8 coil is equal to the downward pulling force (plus gravity force) of the lower figure-8 coil. The vertical position of the vehicle with respect to the center plane of the figure-8 coils automatically maintains the levitation.

Principle of Lateral Guidance The levitation coils facing each other on the opposite walls of U-shaped guideway are interconnected to make a loop. When vehicle moves sideways (e.g., to the right side), superconducting magnets induce current in the figure-8 coils such that it pushes the vehicle to the left side. However, the same movement reduces current in the left side figure-8 coils, which causes the vehicle to move to the left. This process automatically keeps the vehicle located at the center of the guideway.

Principle of Propulsion Repulsive and attractive forces induced between the propulsion coils on the guideway and the superconducting magnets on the vehicle are used to propel the vehicle. These forces are shown in Figure 10.4. The propulsion coils fixed to both sidewalls of the guideway are energized by a three-phase alternating current from a substation, creating a traveling magnetic field on the guideway. The on-board superconducting magnets are locked to this field and are dragged by the traveling field to propel the Maglev vehicle.

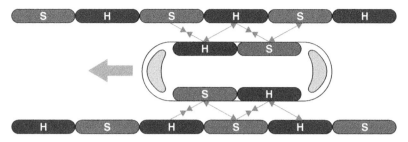

Figure 10.4   Propulsion coils for EDS Maglev

Superconducting Magnets   The original design of the JR-Maglev employed NbTi superconducting magnets [3,4,5,6]. Each magnet is of racetrack configuration with 570-mm straight side and 500-mm diameter arc. The magnets are cooled with LHe pool at 4.2 K. More recently magnets with HTS 1G wire have been tested on a vehicle [7]. An HTS magnet systems consisting of four persistent current HTS coils was developed. The HTS coils are installed in a cryostat and are cooled to approximately 15 K by two-stage GM-type pulse-tube cryocoolers. The HTS magnet is operated in a persistent current mode at a rated magneto-motive force of 750 kA (the same as on the original NbTi coils). The running tests were conducted on the Yamanashi Maglev Test Line, with a top speed of 553 km/h achieved on December 2, 2005. The test result demonstrated that the HTS coils generated no excessive vibration or heat load.

10.2.2   Electromagnetic Suspension (EMS)

Electromagnetic suspension (EMS) employs actively controlled electromagnets (on the vehicle) attracted to a magnetic iron rail fixed to T-shaped guideway [8a]. Both levitation and lateral guidance is achieved without mechanical contact between the vehicle and guideway. In the Transrapid Maglev system, levitation is achieved with attractive forces between levitation electromagnets in the vehicle and ferromagnetic steel reaction rails attached to the guideway as shown in Figure 10.5. The levitation magnets attract the vehicle to the guideway from below, and the guidance magnets keep it centered on the guideway. An electronic system controls excitation in electromagnets actively assures a uniform distance of 10 mm between the magnets and the rails. A synchronous three-phase linear motor, embedded in reaction rails, provides the propulsion power without contact between the vehicle and guideway. Power for levitation, guidance, and hotel loads is transferred

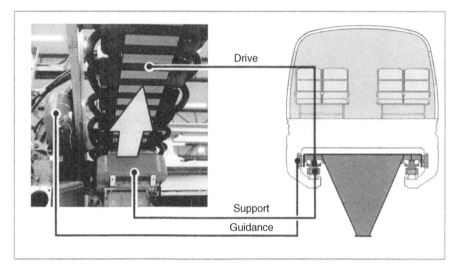

Figure 10.5 Transrapid Maglev concept of electromagnetic suspension (EMD) (Courtesy of ThyssenKrupp)

Figure 10.6 Levitation pole assembly of the Transrapid Maglev (Courtesy of ThyssenKrupp)

from rails to the vehicles through a linear generator without physical contacts.

Principle of Magnetic Levitation Levitation is achieved with a 10-pole assembly shown in Figure 10.6. These poles employ room-temperature copper windings. The gap between magnet poles and the rails is monitored constantly. The DC excitation current in the magnet poles is actively controlled with an electronic system. The current is increased when the gap tends to open and is decreased when the gap tends to close. This process maintains a constant gap of 10 mm between the magnet poles and the rails.

Principle of Lateral Guidance   Lateral guidance is achieved by attraction between the electromagnets mounted on a vehicle and fer-romagnetic reaction plates mounted on the side of guideway, as shown in Figure 10.5. Just like the levitation magnets, the current in these guidance magnets is also controlled electronically as a function of gap length between the magnets and the reaction plate on the guideway.

Principle of Propulsion   The rails, to which levitation poles are attracted, are made of laminated steel similar to that employed in elec-tric motors. A linear three-phase propulsion winding is housed in slots cut into the rails, as shown in Figure 10.5. When supplied with a three-phase current, a magnetic field is created that travels lengthwise along the rail. The combination of levitation poles and propulsion windings in the rail form a linear synchronous motor. The poles of the levitation magnets lock with this field and are pulled along the rail. The speed of the vehicle is proportional to the AC frequency in the traction winding. The frequency is varied for achieving the desired speed. The AC wind-ings are constructed by simply looping a stranded insulated cable in the rail slots, as shown Figure 10.7.

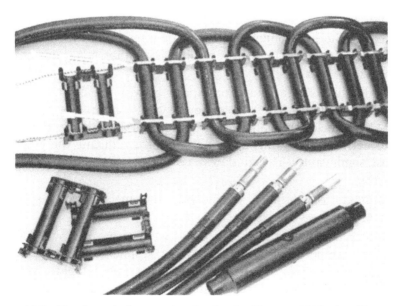

Figure 10.7   Traction winding construction for the Transrapid Maglev (Courtesy of ThyssenKrupp)

Role of Superconducting Magnets The Transrapid system currently in operation in Shanghai and Emsland, Germany, employs non-superconducting levitation coils made from copper. This system has the following issues:

- The Transrapid is noisy at high speeds because the vehicles follow the track with an 8-mm gap, which transfers the uneven movement of the track into vehicle.
- The small gap is filled with harmonic fields created by the AC propulsion winding. The interaction of these fields with the levitation magnets creates harmonic forces, which are felt as vibrations. However, the harmonic fields coupling to the vehicle coils are also utilized for powering the on-vehicle systems.
- A secondary suspension system (mechanical and heavy) is employed to damp out these vibration, but it is inadequate. The Transrapid ride at speed of about 300 km/hr feels like driving a car on cobble road. The ride at high speeds is noisy and uncomfortable.

One solution to all these problems is to employ HTS technology:

- Open the air gap to 50 mm by replacing copper coils with HTS coils.
- Reduce the harmonic magnetic fields interacting with suspension magnets in the gap to a negligible level that generates smaller harmonic forces.
- Guide the vehicle in inertial space by varying current in HTS magnets at 1 kHz. This way guideway variations are not transferred to the vehicle. The vehicle can operate on a guideway that is not perfectly smooth—this has a potential for reducing the guideway cost significantly. The guideway cost is the lion share of Maglev system.
- Eliminate or simplify significantly the mechanical suspension system on the vehicles.
- Reduce significantly the vehicle weight by removal of the heavy copper coils, their power supplies, and mechanical suspension system. The net result of this would be lighter vehicles, reduced fuel consumption, and a lighter and less expensive guideway.

10.3   DESIGN ANALYSIS

This section describes basic principles for designing the Maglev EDS and EMS systems for high-speed applications.

10.3.1   Electrodynamic Suspension Maglev

Many arrangements have been proposed [8] for EDS systems, but the one employing the figure-8 coil arrangement used in the JR-Maglev [3] system is discussed here. This EDS Maglev system employs separate levitation/guidance and propulsion coils, all mounted on the guideway as shown in Figure 10.2. The superconducting coils are located on the vehicle. All coils are of the air-core type, meaning there is no ferromagnetic material in the magnetic circuit. To calculate interaction among these coils, it is necessary to calculate self-inductance and mutual inductance among all coils. Because of the complex nature of the coil arrangement and the difficulty of calculating parameters without the assistance of 3D finite-element programs, only the analysis procedures are discussed in this section for levitation/guidance and propulsion systems.

Levitation/Guidance   System   An   arrangement   of   levitation/ guidance   coils   and   a   superconducting   coil   is   shown   in   Figure   10.8. Superconducting magnets are employed to levitate, guide, propel, and

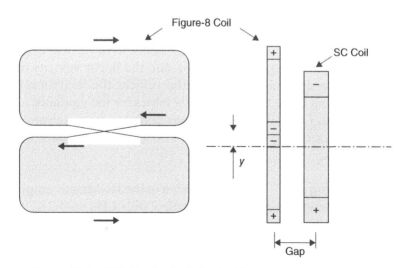

Figure 10.8   EDS Maglev levitation/guidance magnet arrangement

brake the vehicles. Levitation is created by the magnetic repulsion between the on-vehicle superconducting coils (racetrack design) and the magnetic fields induced in the nonsuperconducting track coils (figure configuration) located on the sides of the U-shaped track. The figure-8 coils provide stability to the vehicle's motion by balancing against both up and down motion. At equilibrium, the superconducting coils pass the figure-8 coils at the midpoint or crossover of the figure 8. No net circulating current is induced in the figure-8 coils in this case. If the vehicle falls below the midpoint, the induced currents are set up in the track coils, creating a repelling force from the bottom half of the figure-8 loop, and an attractive force from the top half of the loop (and vice versa, if the vehicle were to rise above the equilibrium point). Currents are induced in the track coils by the moving superconducting coils along the guideway, and levitation begins after the vehicle reaches a speed of about 100 km/hr. At lower speeds, the vehicles travel on wheels. Guidance is provided by the same magnetic forces that repel the vehicle from the track coils. Since the track coils are located on the sides of the track, the induced currents from the sideway motion repel the vehicle if the vehicle moves closer to one of the sides.

Circuit modeling is one of the ways to analyze this EDS system. This requires calculations for the self-inductance and mutual inductance for all the participating coils, including vertical, lateral, and linear (along guideway) displacements. It is difficult to develop close form formulas for the various inductances. However, it is possible to employ 3D finite-element codes for calculating these quantities. For example, tables of the levitation force due to the interaction between a superconducting coil and a figure-8 coil could be created as functions of vertical displacement y and the lateral gap (Figure 10.8), and the linear velocity of the vehicle. During a dynamic analysis of the vehicle, the levitation force could be read from these tables. Similarly tables for the guidance forces could be created. Once design of various coils has been selected, the above-said tables could also be created to include test data on the model coils.

Superconducting Coil The initial design of the JR-Maglev employed LTS NbTi superconducting magnets cooled with LHe to 4.2 K. These DC magnets were designed to operate in persistent mode. However, during operation the motions of the vehicle induced low frequency AC in the superconducting winding. These AC currents generated losses in the superconducting windings and resulted in a decay of the persistent current, and some cases caused a magnet quench.

To solve this problem, Japan Railways decided to evaluate the HTS magnets employing Bi-2223 tapes operating at 20 K. The purpose was to take advantage of HTS wire's tolerance to much larger local loss and temperature variation than NbTi. Test coils were built and tested as described by Igarashi [3]. Since the objective was to operate the HTS coils in persistent mode too, it was important to know how well they performed. On testing of an HTS prototype coil, only a small current decay rate was observed (0.44%/day while cooled with a cryocooler below 20 K.) This was a very encouraging outcome. The HTS coil consisted of 12 single-pancake coils, which were wound with four parallel Ag-sheathed Bi2223 tapes. The HTS coil was connected to a persistent current switch made of an YBCO thin film, and cooled by a G-M (Gifford-McMahon) type two-stage pulse-tube cryocooler. Detachable current leads were used to reduce heat leakage to the first stage of the cryocooler. These tests confirmed that the HTS magnets could operate in persistent mode for day. In a real-life operation these magnets could be recharged to their rated current at the end of each day.

Propulsion Coils   Racetrack shape propulsion coils are mounted on the sides of the guideway, as shown in Figure 10.1. Each coil is a full-pitch (λ) coil, meaning the center-to-center width of a coil is equal to the pole-pitch of the superconducting coils on the vehicle. These propulsion coils are arranged in two layers as shown in Figure 10.9. A magnetic wave traveling along the track is created when the propulsion coils are energized with three phase currents. The vehicle is pulled along the guideway by linear synchronous motor action created by the interlocking action between the traveling wave and the superconducting coils. The propulsion coils are energized from a power source of variable frequency that is available along the length of the guideway.

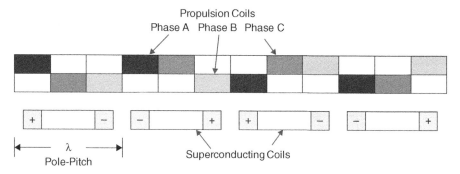

Figure 10.9   Arrangement of propulsion and superconducting coils

The velocity of the vehicle is proportional to the frequency of the current in the propulsion coils. To minimize losses, the propulsion coils are energized in small sections through which a vehicle is passing at the time.

The coil cross section and the ampere-turns in each coil are determined based on the traction power need and the coupling between the propulsion coils and superconducting coils. Propulsion coils are made from copper or aluminum using conventional coil manufacturing technology. To minimize AC losses, the coil conductor should be made of fully transposed strand construction (i.e., Litz wire). Ambient air cools these coils.

10.3.2   Electromagnetic Suspension Maglev

The EMS Maglev system of the Transrapid (Figure 10.5) is used for discussion in this section. This system employs only normal conducting copper coils and windings with ferromagnetic materials in the magnetic circuits. The levitation and propulsion functions are accomplished with one set of magnets. Lateral guidance is achieved with a separate magnet system. An arrangement of these magnets for the Transrapid vehicle is shown in Figure 10.10. The magnetic features of these functions are discussed in this section.

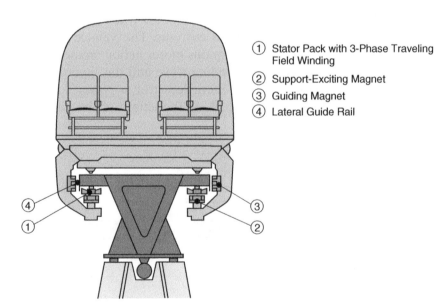

① Stator Pack with 3-Phase Traveling
   Field Winding

② Support-Exciting Magnet

③ Guiding Magnet

④ Lateral Guide Rail

Figure 10.10   Arrangement of the magnet systems for a Transrapid vehicle (Courtesy of ThyssenKrupp)

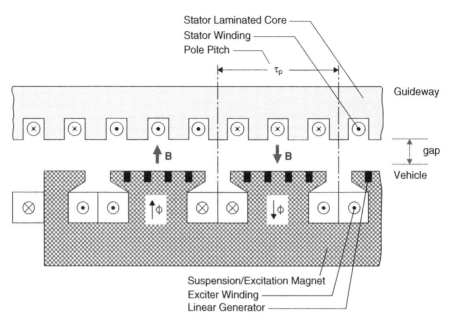

Stator Laminated Core

Stator Winding

Pole Pitch

τ_p

Guideway

B   B

gap

Vehicle

Suspension/Excitation Magnet

Exciter Winding

Linear Generator

Figure 10.11 Transrapid levitation/propulsion magnet system (Courtesy of ThyssenKrupp)

Levitation System An arrangement of levitation magnets facing a reaction rail [8a] is shown in Figure 10.11. The Transrapid vehicle is levitated by balancing the attractive force between the levitation poles and reaction rails against the gravity force. A gap length of 8 mm is maintained between the magnet poles and the rail at all times, both in the stationary and moving stages. According to Earnshaw's theorem, any combination of static magnets cannot be in a stable equilibrium. However, it is possible to achieve stable operation by constantly varying the magnet excitation. The air-gap length and levitation magnet acceleration signals are utilized for controlling the current supplied to the magnet windings. The attractive force between magnets and the reaction rails is a function of air-gap field, B as shown in Figure 10.11. This levitation force (F_l) is given [9] by

$$F_l = 0.5w\tau\left(\frac{B_m^2}{2\mu_0}\right)N_p, \qquad (10.1)$$

where

B_m = peak of the sinusoidal air-gap field,
μ_0 = permeability of free space,
N_p = number of poles,
w = width of reaction rail,
τ = pole pitch.

Due to the leakage field, the value of B is not uniform. A simple 2D finite-element program could be employed to obtain the field distribution in the air gap. The total force per pole is obtained by integrating equation (10.1) over the pole surface area (pole-pitch × pole depth).

The Transrapid employs 10-pole assemblies (Figure 10.6) for levitation and propulsion. All poles carry the same current and are controlled together as a single assembly.

Propulsion System   Once an EMS vehicle is levitated, it could be propelled with a linear induction or synchronous motor. However, the Transrapid system utilizes levitation magnets for propelling the vehicle too. The reaction rails are made of laminated steel and have slots to house a three-phase armature winding employing one slot/pole/phase [10]. The combination of levitation magnets and the armature winding in the rail creates a linear synchronous motor as shown in Figure 10.11. The armature winding is created by placing an insulated cable in the slots using the wave pattern shown in Figure 10.7. To the first order, the propulsion force (per pole) can be obtained from

$$F_t = 1.5 B_m I_m q w \eta k_d N_p, \tag{10.2}$$

where

F_t = propulsion force,
I_m = peak of the sinusoidal current in propulsion winding,
q = slots/pole/phase in propulsion winding,
k_d = distribution factor,
η = efficiency of propulsion motor (~98%).

The poles needed to support a vehicle are given by equation (10.3), which is done by eliminating B_m from equations (10.1) and (10.2):

$$N_p = \frac{F_t^2}{F_l} \frac{\tau}{I_m^2 q^2 9\mu_0 w} \frac{1}{\eta^2 k_d^2}. \tag{10.3}$$

The field in the air gap is usually not uniform along the rail width and the rail length. A more complete treatment of an analytical approach for calculating propulsion force is provided by Nasar [11]. However, for a better estimate of the linear force, this calculation should be made with a 3D finite-element program.

The armature winding is divided into sections along the length of the guideway. These sections are energized in a leapfrog way in order to improve the system's efficiency; that is, the section containing the vehicle and the section following in the direction of travel are energized. The armature winding is supplied from ground converters capable of controlling the voltage, current, and frequency.

Guidance System   Lateral guidance is achieved by the attraction between the electromagnets and ferromagnetic plates mounted on the side of the guideway as shown in Figure 10.10. A constant air gap is maintained between each electromagnet and guideway. However, this approach transfers guideway sways into the vehicle, making the ride uncomfortable.

Secondary Suspension System   A high level of vibrations in the Maglev vehicle is experienced because of the variations in the levitation and guidance air gaps and the air-gap harmonic field effects on the propulsion force. A secondary mechanical suspension system is incorporated in each vehicle to damp out these vibrations. Despite this the currently operating Transrapid vehicles are still noisy and uncomfortable at high speeds.

Superconducting Suspension Magnet System   It is possible to eliminate or significantly reduce vibration by increasing the air-gap length to between 40 and 50 mm. The large gap essentially decouples the vehicles from guideway irregularities and reduces the effect of the armature harmonic fields on the propulsion force. It is even possible to eliminate the mechanical secondary suspension system (bulky and heavy). Such a system was developed [12] by Northrop Grumman during the early 1990s. This system is described in detail in Section 10.6.

10.4   CHALLENGES (TECHNICAL/ECONOMIC)

Although both the EDS and EMS Maglev systems look so futuristic and environmentally friendly (energy efficient), they face many technical and economic challenges. Some of these challenges are listed below:

· High capital cost, primarily that of the guideway.
· Right-of-way availability.
· Noncompatibility with existing railroad system.
· Lack of public enthusiasm for its introduction.

These technical challenges from magnetic point of view are discussed here.

10.4.1 EDS System Challenges

The EDS system employs a U-shaped guideway that carries levitation, guidance, and propulsion coils made of copper or aluminum. Because the wall of the guideway must bear the vertical and lateral loads of vehicles, they tend to be thick. These reinforced concrete walls tend to be high in cost due to the long stretch of the guideway. Furthermore the heavy U-shape of the guideway requires heavier and closely spaced vertical support pillars, which are esthetically not very appealing. Then again, the biggest attraction of this Maglev system is its inherent stability and fixed guideway switches.

On the magnetic side, the superconducting magnets are located on vehicles and usually operate in persistent mode to minimize the need for power supply in the vehicles. Because of the very large air gap (~100 mm) between the superconducting and guideway magnets, the superconducting magnets must operate at very high field levels (about 2.5 T). The strong magnetic DC fields onboard the vehicles require heavy shielding to make them accessible to passengers with pacemakers or magnetic data storage media such as hard drives and credit cards.

The JR-Maglev has employed NbTi magnets cooled to 4.2 K for many years and yet always has had magnet stability and other operational problems. Recently prototype magnets using the 1G HTS wire were tested and were found to be suitable for fixing problems that plagued NbTi magnets for a long time. However, HTS wire is currently expensive, and it is not clear if the HTS magnets will make economic sense until the wire price falls below $5 to $10/kA-m. However, JR-Maglev expects this system with a top speed of 650 km/hr to be commercially available by 2025 for service between Tokyo and Nagoya.

10.4.2 EMS System Challenges

An EMS system based on the Transrapid is already in commercial service in Shanghai, China. This system employs a T-shaped guideway with an 8-mm air gap between levitation and guideway reaction rails. Because of this small air-gap length, the guideway must be constructed with tighter tolerance requirements. This adds to the cost of the guideway to make it sufficiently rigid. The Maglev vehicles are also heavy due to the use of iron-core copper magnets and the bulky mechanical secondary support system. Since the Maglev vehicle wraps around the guideway, it becomes essential to employ movable track switches (with limited speed capability, <56 m/s), which adds to the cost of the system.

However, the EMS system employs iron-core magnets, which eliminates or minimizes stray magnetic field inside a vehicle and in surrounding areas. The vehicles are levitated at all speeds and do not require the supplemental wheel support that is necessary for the EDS system.

The EMS Maglev system developed by Grumman employing HTS magnets is similar to the Transrapid system, but it has an air-gap length up to 50 mm. This eliminates vibrations and reduces the cost of the guideway, which could be built to more relaxed tolerance requirements. Furthermore both levitation and guidance functions are performed by the same set of HTS magnets. The vehicles are lighter too because the HTS magnets are lighter, and the secondary mechanical support system has been eliminated or significantly reduced. The lighter vehicles also affect the structural design of the guideway in a positive manner. This system is described in detail in Section 10.6.

10.5 MANUFACTURING ISSUES

The EDS and EMS systems have been under constant development for 25 years or more. Most of the manufacturing issues related to magnet systems have been well addressed. Economic issues relating to the cost of the overall system seem to be the major obstacle to their commercial applications.

10.6 PROTOTYPES

Both EDS and EMS systems are in operation. Improvements are being considered in the JR-Maglev EDS system by replacing NbTi magnets with HTS magnets. It is not clear at what pace the HTS magnets will be introduced.

The Transrapid based on EMS is also addressing issues relating to vibrations at high speeds and fixes are being considered. It is not known what specifically is being planned. However, the author was part of the team [12] that developed an the HTS Maglev concept based on the EMS system. This system is described below.

10.6.1 Northrop Grumman Concept

During the early 1990s Northrop Grumman (NG) developed [12] an EMS Maglev concept with support from the United States Army Corps of Engineers. Like the Transrapid system, the NG system employs iron-core levitation magnets but has HTS excitation windings. This system has the characteristics of a low stray field in the vehicle and surrounding areas, a uniform load distribution along the full length of the vehicle, and a small pole-pitch for smoother propulsion and ride comfort. It is also levitated at all speeds and incorporates a wraparound design around the guideway for safer operation. The NG Maglev system has all the advantages of an EMS system identified above, while eliminating (or significantly improving) drawbacks associated with the normal magnet-powered Transrapid system. Improvements include a larger air gap, lighter weight (both for vehicles and guideway), a lower number of control servos, and higher offline switching speeds. The design also incorporates a vehicle tilt (+/– 9°) for a higher coordinated turn and turn-out speed capability.

Levitation, Guidance, and Propulsion System Design A vehicle with iron-core magnets and reaction rails mounted on the guideway is shown in Figure 10.12. The laminated iron-core magnets and reaction iron rails are oriented in an inverted-V configuration, with the attractive forces between the magnets and rail acting through the vehicle's center of gravity (*cg*). The vertical control forces are generated by sensing the air-gap clearance on the left and right side of the vehicle and adjusting currents in the control coils to maintain a relatively large air gap (40 mm) between the iron rail and magnet face. Lateral control is achieved by the differential measurement of the gap clearance between the left and right sides of the vehicle magnets. The corresponding magnet control coils are differentially driven for lateral guidance control. Through this process, the vehicle control with respect to the rails is achieved in the vertical, lateral, pitch, and yaw directions. The control mechanism has been recently described by Gran [13].

Two magnets combined as shown in Figure 10.13, make up a "magnet module." Each magnet module is a C-shaped, laminated iron core with an HTS coil wrapped around the center body of the magnet, and two

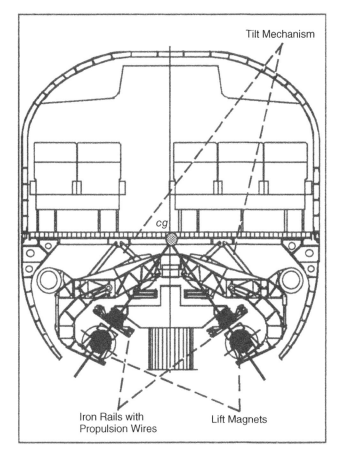

Tilt Mechanism

cg

Iron Rails with
Propulsion Wires

Lift Magnets

Figure 10.12   Cross section of a vehicle showing levitation (lift) magnets, iron rail, guideway, and tilt mechanism (Courtesy of Advanced Energy Systems / Northrop Grumman)

copper control coils wrapped around each leg. Vehicle roll control is achieved by offsetting the magnets by 20 mm in a module to the left and right side of a 200-mm-wide reaction rail. The control is achieved by sensing the vehicle's roll position relative to the rail and differentially driving the offset control coils to correct for roll errors.

The reaction rails are also laminated and contain slots for the installation of a set of three-phase AC linear synchronous motor (LSM) propulsion winding. The winding is powered with a variable-frequency, variable-amplitude current that is synchronized to the vehicle's speed. The speed variations are achieved by increasing or decreasing the frequency of the AC current. The magnet design was optimized using 2D and 3D finite-element codes to ensure that all levitation, guidance, and

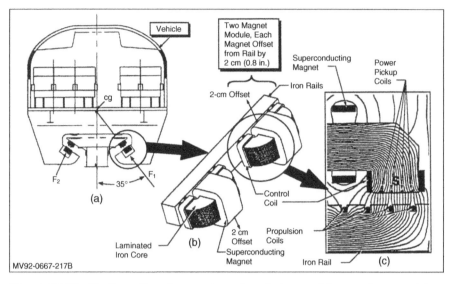

Figure 10.13 Configuration of magnets and control, propulsion, and power pickup coils (Courtesy of Advanced Energy Systems / Northrop Grumman)

propulsion control requirements are met simultaneously. The power pickup coils (for the HTS coils and the hotel load in the vehicles) located in each magnet pole face were designed to operate at all speeds, including zero speed, when the vehicle is standing still [14].

The low magnetic field in the passenger compartment and surrounding areas represents an important characteristic of the Northrop Grumman Maglev system. Figure 10.14 identifies the constant flux densities in the vehicle (cabin) and station platform for the NG Maglev baseline design. The flux density levels above the seat are less than 1 gauss, which is very close to the earth's 0.5 gauss field level. On the platform, the magnetic field level, when the vehicle in the station, does not exceed 5 gauss, which is considered acceptable in hospitals using magnetic resonant imaging (MRI) equipment. The data in Figure 10.14 are based on a 3D magnetic field analysis for the case of no shielding. With a modest amount of shielding, these levels could be further reduced if considered necessary.

Iron-core magnets employing an HTS winding can reduce the magnetic field experienced by the HTS wire to less than 0.35T. With a reasonable operating current density this would enable operation of the HTS 2G wire at 77 K. The HTS coil also experiences only a small mechanical load due to the interaction of leakage magnetic field and the coil current. The HTS coils were powered with a patented constant

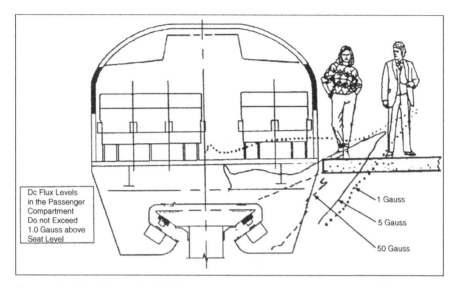

Dc Flux Levels
in the Passenger
Compartment
Do not Exceed
1.0 Gauss above
Seat Level

1 Gauss

5 Gauss

50 Gauss

Figure 10.14 Magnetic field in the cabin and surrounding area (Courtesy of Advanced Energy Systems / Northrop Grumman)

current loop controller [15] that diminishes the rapid current variations on the coil and minimizes the AC losses in it. By employing the CTC cable in the HTS coils, it might even be possible to eliminate the control coils surrounding the iron poles. This possibility should be considered in future evaluations of this design concept.

Northrop Grumman recommended the guideway configuration shown in Figure 10.15. This configuration was determined to be lowest in cost and found to be relatively insensitive to span length [16]. This has an important implication for the guideway that must be installed in areas such as the US interstate highway system, for which the span length would range widely and depend on local road conditions. This system is also visually less intrusive because the single column creates less shadow and is esthetically pleasing.

A 5-degree-of-freedom analysis of the interactive effects of the vehicle traveling over a flexible guideway was also undertaken [13,17]. The guideway irregularities resulting from random step changes, camber variations, a span misalignment, and rail roughness were included in the simulation. Also studied were linearized versions of the vehicle levitation and lateral control loops. The results indicate that passenger comfort levels could be maintained.

In 1995 Northrop Grumman built the test stand shown in Figure 10.16 for testing prototype levitation magnets employing LTS and HTS

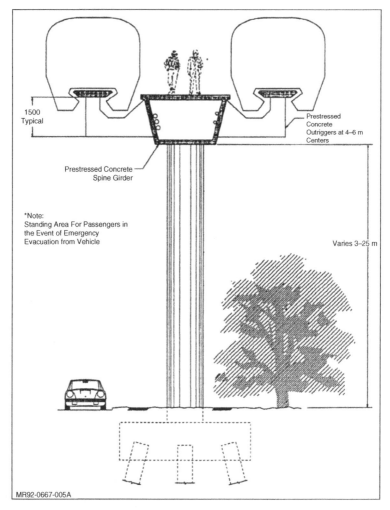

Figure 10.15 Baseline spine girder configuration (Courtesy of Advanced Energy Systems / Northrop Grumman)

superconductors. The suspension magnet shown in Figure 10.17 consisted of a U-shaped laminated iron core, an HTS coil around the center body of the core, and two copper control coils wrapped around each pole. This magnet faced a slotted rail that was fitted with a three-phase winding for testing the propulsion force. The rail supported from actuators could be moved up and down from very slow to high frequency for simulating vibration in an actual operation on a guideway. The magnet assembly was simply suspended using the attractive force between the magnet poles and the rail. The system

Figure 10.16   Test apparatus for testing EMS system employing HTS magnet

was designed for a 30-mm gap between the magnet pole faces and the rail.

The control system shown in Figure 10.18 is used to achieve levitation with the hybrid superconducting/normal coil magnet system. Separate power supplies are provided for the superconducting and normal coils. The supply driving the superconducting coil is a constant current power supply that automatically adjusts its output voltage to maintain a load current constant regardless of its load impedance. The normal coils are driven by a conventional regulated supply whose output current is set by a gap sensor that drives the gap to the desired spacing. A current sensor, included in the normal coil circuit, interfaces with the constant current power supply through a low-pass filter and controls the set point of the constant current power supply driving the superconducting coil.

The operation of the system is as follows: Upon energizing the system, the current set points of both supplies are set to zero. The gap

Figure 10.17 U-Shaped magnet assembly with an HTS coil cooled with cryocooler

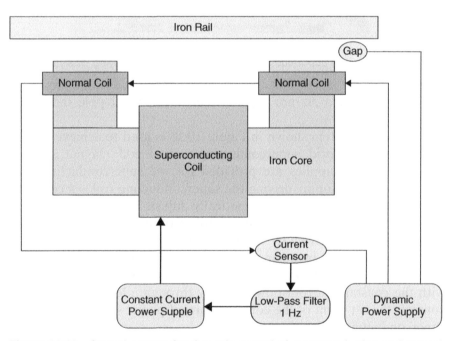

Figure 10.18 Control system for dynamic control of superconducting and normal coils

sensor detects a larger than desired gap and starts the current flow in the normal coils. At the same time the current sensor in the normal coil circuit detects a current flow, which triggers the current flow in the superconducting coil. When sufficient MMF is generated by the copper and superconducting coils the magnet begins to levitate and the gap sensor begins to maintain correct gap. At this point the entire MMF is supplied by the superconducting coil and no current flows in the normal coils. Should a disturbance such as an additional weight be placed on the levitating magnet, the gap would begin to open and the gap sensor would detect the change. This immediate reaction in the normal coil current creates the required levitating force. The current sensor instantaneously detects the current change and begins to increase the current in the superconducting coil. The increased MMF then begins to close the gap, and the detector starts to lower the current in the normal coils until equilibrium is restored and no current flows in the normal coils.

This system was successfully demonstrated. The system as shown in Figure 10.16 can support two people on a suspended magnet while maintaining the design air-gap length of 40 mm. The rail was moved at frequencies of a few Hz to 1 kHz and the magnet remained suspended throughout.

U-shaped magnet assemblies were used for the baseline design of the NG Maglev system, but they were found to have an excessive leakage field. The multiple poles magnet assemblies as used in the Transrapid system have a much lower leakage reactance and can be implemented with superconducting coils. In a multi-pole assembly the superconducting coils are installed around each pole, and all superconducting coils are housed in a common cryostat as explained in the US Patent [18] 5,479,145. The normal coils are placed around each pole just outside the cryostat. This Maglev project was not carried further because of lack of support from the government and company sources.

10.7 SUMMARY

Both the EDS and EMS systems have been sufficiently developed and are ready for commercial deployment contingent on their economic feasibility. Advances in HTS magnet technology will be beneficial to both systems. For the EDS system, it offers more reliable system operation at expense of extra cost. For the EMS system, it has the potential for reducing weight, increasing efficiency, and providing smoother vibration free ride.

REFERENCES

1. H. W. Lee, K. C. Kim, and J. Lee, "Review of Maglev Train Technologies," *IEEE Trans. Magnetics* 42(7): 1917–1925, 2006.

2. R. Thornton, ed., "Special Issue on Linear Motor Powered Transportation," *Proc. IEEE* 97(11): 2009.

3. M. Igarashi, H. Nakao, M. Terai, T. Kuriyama, S. Hanai, T. Yamashita, and M. Yamaji, "Persistent Current HTS Magnet Cooled by Cryocooler (1)— Project Overview," *IEEE Trans. Appl. Superconductivity* 15(2, Pt. 2): 1469– 1472, 2005. DOI 10.1109/TASC.2005.849130.

4. S. Kusada, M. Igarashi, K. Kuwano, K. Nemoto, S. Hirano, T. Okutomi, M. Terai, T. Kuriyama, K. Tasaki, T. Tosaka, K. Marukawa, S. Hanai, T. Yamashita, Y. Yanase, H. Nakao, and M. Yamaji, "Persistent Current HTS Magnet Cooled by Cryocooler (2)—Magnet Configuration and Persistent Current Operation Test," *IEEE Trans. Appl. Superconductivity* 15(2, Pt. 2): 2285–2288, 2005. DOI 10.1109/TASC.2005.849632.

5. K. Tasaki, T. Tosaka, K. Marukawa, T. Kuriyama, S. Hanai, M. Yamaji, K. Kuwano, M. Igarashi, K. Nemoto, S. Hirano, T. Okutomi, H. Nakao, and M. Terai, "Persistent Current HTS Magnet Cooled by Cryocooler (3)—HTS Magnet Characteristics," *IEEE Trans. Appl. Superconductivity* 15(2, Pt. 2): 2289–2292, 2005. DOI 10.1109/TASC.2005.849633.

6. T. Tosaka, K. Tasaki, K. Marukawa, T. Kuriyama, H. Nakao, M. Yamaji, K. Kuwano, M. Igarashi, K. Nemoto, and M. Terai, "Persistent Current HTS Magnet Cooled by Cryocooler (4)—Persistent Current Switch Characteristics," *IEEE Trans. Appl. Superconductivity* 15(2, Pt. 2): 2293– 2296, 2005. DOI 10.1109/TASC.2005.849634.

7. K. Kuwano, M. Igarashi, S. Kusada, K. Nemoto, T. Okutomi, S. Hirano, T. Tominaga, M. Terai, T. Kuriyama, K. Tasaki, T. Tosaka, K. Marukawa, S. Hanai, T. Yamashita, Y. Yanase, H. Nakao, and M. Yamaji, "The Running Tests of the Superconducting Maglev Using the HTS Magnet," *IEEE Trans. Appl. Superconductivity* 17(2, Pt. 2): 2125–2128, 2005. DOI 10.1109/ TASC.2007.899003.

8. R. G. Rhodes and B. E. Mulhall, *Magnetic Levitation for Rail Transport*, Clarendon Press, Oxford, 1981.

8a. K. Heinrich, and R. Kretzschmar, "Transrapid Maglev System," Hestra-Verlag Darmstadt 1989, ISBN 3-7771-0209-1.

9. S. S. Kalsi, R. Herbermann, C. Falkowski, M. Hennessy, and A. Bourdillon, "Magnet Design Optimization for Grumman Maglev Concept," 1993 Maglev Conference, Argonne National Laboratory, Paper PS2-7, May 19–21, 1993.

10. C. P. Parsch and H. G. Raschbichler, "The Iron Core Long Stator Synchronous Motor for the Emsland Test Facility," *ZEV-Glas. Annalen* 5(7–8): 225–232, 1981 (in German).

11. S. A. Nasar and I. Boldea, *Linear Electric Motors: Theory, Design and Practical Applications*, Prentice-Hall, Englewood Cliffs, NJ, 1987.

12. M. Proise, L. Deutsch, R. Gran, R. Herbermann, S. S. Kalsi, and P. Shaw, "System Concept Definition of the Grumman Superconducting Electromagnetic Suspension (EMS) Maglev Design," 1993 Maglev Conference, Argonne National Laboratory, Paper OS4-4, May 19–21, 1993.

13. R. Gran, "Optimizing the Controller Design to Guide the Motion of a Maglev Train," Maplesoft Application Demo, Waterloo Maple Inc., 2008.

14. S. S. Kalsi, "On-vehicle Power Generation at all Speeds for EMS Maglev Concept," 1993 Maglev Conference, Argonne National Laboratory, Paper PS5-12, May 19–21, 1993.

15. R. Herbermann, "Self Nulling Maglev Suspension System," 1993 Maglev Conference, Argonne National Laboratory, Paper PS1-7, May 19–21, 1993.

16. B. Bohlke and D. Burg, "Parametric Design and Cost Analysis for EMS Maglev Guideway," 1993 Maglev Conference, Argonne National Laboratory, Paper PS3-2, May 19–21, 1993.

17. R. Gran and M. Proise, "Five Degree of Freedom Analysis of the Grumman Superconducting Electromagnetic Maglev Vehicle Control/Guideway Interaction," 1993 Maglev Conference, Argonne National Laboratory, Paper PS4-6, May 19–21, 1993.

18. S. S. Kalsi, "Superconducting Electromagnet for Levitation and Propulsion of a Maglev Vehicle," US Patent 5,479,145 issued December 26, 1995.

11

MAGNET APPLICATIONS

11.1 INTRODUCTION

Large electrical magnets currently are used in a variety of industrial and military settings. The applications range from medical uses to process manufacturing and purification to scientific research. The manufacture of such magnets with HTS materials is an area of research receiving much interest. The properties of HTS materials such as BSCCO-2212, BSCCO-2223, and YBCO-123 are very attractive at low temperatures (<20 K) and YBCO-123-coated conductors available commercially now are suitable for making useful devices cooled at higher temperature of liquid nitrogen (~77 K).

Traditionally these magnets employed copper wire or LTS. However, magnets based on HTS wire compare favorably to conventional copper magnets for a number of reasons:

Smaller and lighter. The high current density of the HTS wire enables designers to reduce the size and weight of coils in magnet applications by as much as 40% to 80% compared to those made with copper wire.

More efficient. Because HTS wire has virtually no resistance to direct currents, HTS coils and magnets have higher energy efficiency and lower operating cost than copper based systems.

Applications of High Temperature Superconductors to Electric Power Equipment,
by Swarn Singh Kalsi
Copyright © 2011 Institute of Electrical and Electronics Engineers

Higher magnet fields. The high current density in HTS wires enables higher magnetic field magnets.

Greater thermal stability. Coils and magnets using HTS wire operate in a stable "cold environment." These magnets operate within a narrow temperature range because of the very low level of heat generated by HTS magnets. Conversely, copper coils generate high amounts of heat, making it difficult to maintain proper temperature when the magnet application requires tight thermal control.

Longer magnet life. The "cold cryogenic environment" within which HTS magnets operate eliminates a common source of product failure, namely heat. Thermal cycling, which shortens the useful lifespan of copper wire coils and magnets, is not a concern with HTS magnets.

Easier cooling. The commercial availability of HTS conductors has provided an incentive to design and build larger and more ambitious devices with performance advantages when compared to low-temperature superconducting (LTS) devices built with NbTi and Nb3Sn superconductors operating at liquid helium temperature. In particular, steady state AC operation is extremely difficult for conduction-cooled LTS magnets because poor thermal conductivity at low temperatures can lead to unacceptable thermal gradients and magnet quenching.

Current Gifford–McMahon (G-M) refrigerators can provide only about 1 W of refrigeration at 4 to 5 K; devices built with HTS operate at higher temperatures (20 K and above), where the increased thermal conductivity and specific heat mitigate stability issues and where economical single stage GM refrigerators can provide tens of watts of refrigeration. It is thus possible for HTS magnets to accommodate larger heat loads due to AC harmonics or due to faster current ramps. The higher temperature operation also simplifies the cryostat design and reduces the cryostat cost.

High temperature superconducting magnets are commercially available today. Experience with the initial applications, which were in the military and scientific domains, is expected to give rise to expanded applications addressing other markets. The continuing development of the YBCO-123-coated conductor presently provides material with sufficient critical current density to enable a broad variety of applications.

Many HTS magnets have been built using the BSCCO-2212, BSCCO-2223, and YBCO-123 wires [1,2,3,4,5,6,7,8,9,10,11], and more are under construction. Although the feasibility of building HTS magnets has been well established, their commercial viability is less attractive due to the high cost of HTS wire and the cryogenic cooling system. In the sections that follow a few examples are discussed of such magnets built in recent past and a few examples of those currently under construction.

11.2   AIR-CORE MAGNETS

AMSC [1], Sumitomo [3], and a number of other manufacturers have built conduction-cooled HTS magnets operating in the 20 to 30 K temperature range. These systems offer the advantages of high operational stability and the ability to ramp very quickly. The higher cost and lower performance of HTS material at 20 K compared to LTS material at 4 K is limiting the commercial exploitation. The commercial availability of HTS materials has provided the incentive to design and build larger and more ambitious devices with performance advantages when compared to LTS devices built with NbTi and Nb_3Sn superconductors operating at the liquid helium temperature.

The section below describes 7 to 8 T conduction cooled magnets built with the HTS technology.

11.2.1   High-Field Magnets

AMSC High-Field Magnet   A 7.25-T laboratory magnet [1] utilizing the BSCCO-2223 conductor was built for the Naval Research Laboratory in 1998. Operating at 21 K at full field, the magnet provided field homogeneity of ±1% in a 2-inch warm bore. The system was conduction cooled with a pair of Leybold single-stage cryocoolers that allow cool-down in less than 36 hours and allow extended fast-ramp operation. The HTS current leads, employed in the magnet, facilitated the operation with a total refrigerator input power of 6 kW. The fully integrated system consisted of the magnet, cryogenic system, control and protection system and power supply. Figure 11.1 shows the magnet system.

Table 11.1 summarizes the major features of the magnet. It generated a magnetic field > 7 T in a 2-inch warm bore. The field homogeneity of 1% was specified within a 2-inch diameter spherical volume (DSV) and 2% within a 2-inch diameter cylindrical and 2-inch tall volume. The magnet's HTS winding is conduction cooled and operates at about 25 K. The magnet employs HTS current leads between the 25 K

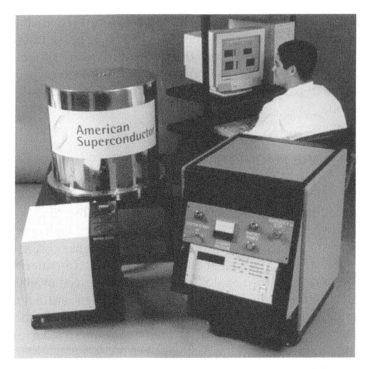

Figure 11.1 Conduction-cooled 7.25-T magnet built by AMSC (Courtesy of American Superconductor Corporation)

magnet and a 45 K intercept temperature. The conduction-cooled leads are employed between 45 K and room temperature. Two single-stage Leybold cryocoolers driven from a single compressor provide refrigeration at the two temperatures.

The magnet was capable of ramping between zero and full field in 200 seconds on a continuous basis. It was powered with a 4-quadrant power supply to permit seamless operation between +7.25 and −7.25 T. A suitable protection system protected it against abrupt quenches and any other unintended operational modes. All components of the magnet system could withstand a shock loading of 6 G in any direction. The magnet cryostat supported experimental equipment weighing up to 250 lbs.

This magnet demonstrated that high field magnets operating above 20 K could be built using the BSCCO-2223 conductor. It also demonstrated that this material provides significant performance advantages for fast-ramp magnets or magnets that require high external heat loads.

Table 11.1   AMSC conduction-cooled 7.25-T magnet features

Parameter	Unit	Value
Peak field in the bore	T	>7
Field homogeneity	%	1
Useful field volume at room-temp.		
• Diameter	inch	2
• Length	inch	2
Room-temperature bore diameter	inch	2
Operating temperature	K	25
Cooling method		Conduction
Ramp time, zero to full-field	s	240
Experiment weight	lb	250
Shock load withstand capability	G	6
Power supply		4 quadrarts

Sumitomo High-Field Magnet   Sumitomo Electric Industries (SEI) of Japan also built an 8-Ta conduction-cooled magnet with the BSCCO-2223 wire as shown in Figure 11.2. Their proprietary process has enhanced the performance of BSCCO-2223 wire significantly, which is called DI-BSCCO (Dynamically Innovative Bismuth-Based HTS wire). A conduction-cooled high-field HTS magnet built with DI-BSCCO with a room-temperature 200-mm diameter bore was tested up to 8.1 T. The design study showed that the higher magnetic field (15 T) could be achieved within about the same envelope of the above-said magnet. This magnet could be used in various industries such as biomedicine, semiconductor, and environmental industries. Table 11.2 lists the key features of this magnet.

11.2.2   Low-Field Magnets

Many low-field magnets have been built for applications such as mine-sweeping, magnetic energy storage, and magnetic separation. Some of these applications are described below.

Minesweeping Magnets   The US Navy funded AMSC to conduct a feasibility study of utilizing HTS conductors in an advanced lightweight influence sweep system (ALISS) magnet. The study demonstrated the feasibility of constructing a 5-MA-m^2 magnet with HTS conductors. This magnet operated at around 20 K and was conduction cooled with a cryogen free closed cycle cryocooler. Since the HTS conductors exhibit a slow transition from superconducting to normal state, they

8T-200mm Magnet

Center Field	8 Tesla
Room temperature bore	200 mm
Inductance	37.7 H

Figure 11.2 Conduction-cooled 8-T magnet built by Sumitomo (Courtesy of Sumitomo Electric Industries)

Table 11.2 Sumitomo conduction-cooled 8-T magnet features

Cooling Method	Cryocooler Conduction Cooling	Cryocooler Conduction Cooling
Maximum magnetic field	8T	15T (example)
Bore diameter at room temperature	200 mm	200 mm
Magnet vessel size (width × depth × height)	900 × 600 × 540 mm	1000 × 700 × 600 mm
Weight	300 kg	500 kg

can tolerate large excursions in local temperatures of the coil without causing the abrupt quench normally experienced in the low-temperature NbTi and Nb$_3$Sn coils. Although a conduction-cooled NbTi magnet operating at 5 K could produce the required dipole moments, because of the small thermal margin (difference between operating temperature and critical temperature of the superconductor) it is liable to quenching from mechanical shocks or from stress relief in the epoxy. These magnets have also practically zero current modulation capability,

Figure 11.3 ALISS minesweeping magnet (Courtesy of American Superconductor Corporation)

again because of their small thermal margin. Magnets made from Nb_3Sn can operate with a cryocooler and provide an additional thermal margin when operated at a moderate current density of 7 to 10 K. However, HTS magnets offer the possibility of much higher thermal margins (over 2000 times that of LTS materials when operated at 20 to 40 K) but are more expensive and less technologically mature than either NbTi or Nb_3Sn. An additional advantage of HTS magnets is the ability to operate in an AC field. The ALISS magnet system required small current modulations (~10% of the nominal current) at frequencies up to 10 Hz. An HTS magnet built under this program absorbed losses generated by such current oscillations, but this would be impossible for a conduction-cooled NbTi coil operating at 5 K.

Figure 11.3 shows an AMSC constructed demonstration HTS minesweeping magnet [11] for airborne superconducting mine countermeasure system in December 1999. The magnet produced a magnet dipole moment of $0.015 MA\text{-}m^2$. The HTS magnet, 0.46 m in diameter and 0.9 m in length, operated at a nominal temperature of 35 K. The magnet was housed in a 560-mm-diameter by 1.5-m-long stainless steel vacuum vessel. The HTS magnet delivered to the US Navy was conduction-cooled and fully integrated with a cryocooler, cryostat, current leads, and ancillary hardware. The magnet was successfully ramped up to 400 A at around 35 K. The magnet current was kept at 400 A for six hours without any voltage runaway. The magnet successfully operated with 50% AC modulation (200 to 400 A) with a trapezoidal current waveform.

Energy Storage   Superconducting magnetic energy storage (SMES) magnets are considered for improving the power quality of process plants or the electric grid, where a short interruption of power can lead to a long and costly shutdown. The poor power quality of a given system could be due to voltage fluctuations, undesirable harmonics introduced by other system loads, and momentary interruption or voltage sags due to faults in the power system.

In the past many developers conceptualized the SMES systems in various sizes using NbTi and Nb3Sn superconductors operating at around 4.2 K. AMSC offered such NbTi-SMES magnets commercially for electric grid stabilization during the 1990s. However, compared to simple electronic inverter solutions these systems were uneconomical. An ability to charge/discharge rapidly is a key requirement for an SMES magnet, but the LTS magnets have only a very limited capability in this area.

To understand the capability of an HTS SMES magnet, a 5 kJ conduction-cooled magnet [8] was built and tested. The magnet system shown in Figure 11.4 consisted of a solenoid coil constructed from a BSCCO-2223 conductor. The coil was epoxy impregnated and cooled with single-stage G-M cryocooler for operation at 100 A (DC) with a substantial AC component created by frequent variation of the current (ramp-up and ramp-down). The dominant heat load in the magnet was due to eddy-current heating caused by the current-ramping operation. The magnet was capable of being charged from zero to full 100 A in 2 seconds and back to zero current in 2 seconds. One hundred such ramp-up/ramp-down cycles could be accommodated before the magnet exceeded the allowable temperature rise.

The successful operation of this magnet demonstrated the feasibility of a conduction-cooled HTS SMES magnet that could withstand rapid charge/discharge cycles. However, commercial HTS SMES systems never emerged, primarily because of the high cost of the HTS conductor and the short persistent current mode due to HTS internal conduction losses.

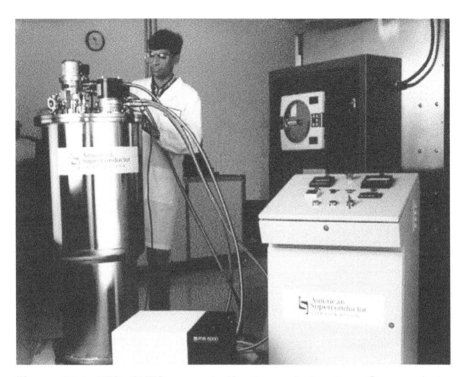

Figure 11.4 HTS SMES magnet (Courtesy of American Superconductor Corporation)

Magnetic Separation   A magnetic separator uses magnetic force to separate materials of various compositions based on their magnetic properties. Usually a magnetic field between 1 and 2T is preferred. Such fields are difficult to generate with copper coils due to high I^2R losses. However, HTS coils can easily create high fields and larger magnetic field gradients, which result in improved performance of the separators. Better separator performance enables the separation of larger amounts of material in a shorter time span, or the separation of more dilute materials. HTS magnetic separators have a variety of industrial applications, notably in the pharmaceutical, environmental, and chemical fields. They process ores, solid wastes, waste gases, and isotope separations and water treatment. Magnetic separators currently purify kaolin clay, a material used in high-quality paper. An HTS separator prototype was used for purifying commercial-grade kaolin clay slurries, and it performed as well as conventional separators, while using less power.

AMSC built an HTS magnet for a prototype separator shown in Figure 11.5. The magnet made of BSCCO-2223 conductor generated a

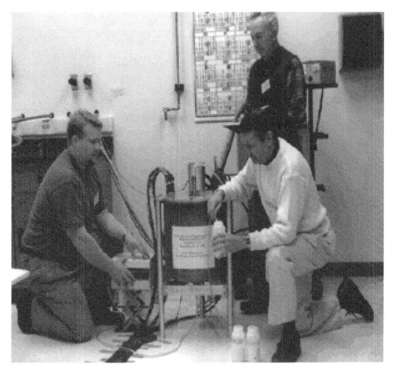

Figure 11.5   HTS magnet for a separator (Courtesy of American Superconductor Corporation)

1-T field and had the overall dimensions of 180-mm outer diameter (OD), 155-mm height, and 50-mm inner diameter (ID). The HTS current leads, with improved shock resistance capability, were employed with the warm end at 75 K and the cold end at 27 K. A two-stage Gifford-McMahon cryocooler cooled the magnet system. The Los Alamos National Laboratory (LANL) successfully tested the separator system in early 1997.

11.3   IRON-CORE MAGNETS

Research in physics and the medical field employs a variety of iron-core magnets. The magnetic fields applied range from a fraction of a tesla to more than 16 T. This section describes a few magnet systems where HTS coils have been recently employed or could be good candidates.

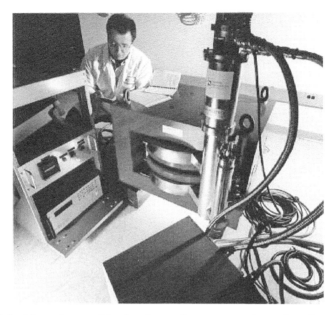

Figure 11.6 Magnet assembly showing major components such as iron core, HTS coil cryostat, cryocooler, and power supply (Courtesy of American Superconductor Corporation)

11.3.1 Beam Bending

An H-shape magnet, consisting of a room-temperature iron magnetic circuit, was built by AMSC [2] for generating a uniform magnetic field of 0.72 T in the air gap between two iron poles. Figure 11.6 shows the magnet system employing BSCCO-2223 HTS coils conduction cooled with a single-stage G-M type cryocooler. The magnet system was designed for steady-state operation for long periods. It was factory tested in the fall of 1996, prior to shipping to a customer in New Zealand. This magnet operated successfully for many years at the Institute of Geological and Nuclear Science (IGNS) in New Zealand transferring an ion beam among three experiment stations.

11.3.2 Induction Heating

Conventional AC induction heating has been used in industry since the 1920s. In 1990 a new concept emerged [12] for DC induction heating using strong electromagnets. The magnet-wire and motor drive technologies available at the time, however, did not permit an economical

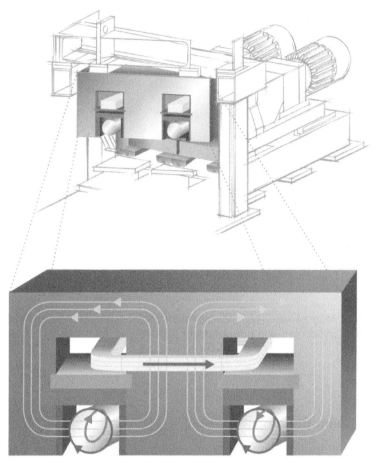

Figure 11.7   Induction heater concept with HTS excitation coil (Courtesy of Zenergy Power)

embodiment of the concept. With the emergence of both HTS as a commercially available conductor and advances in solid-state electric motor drive equipment, this almost 20-year-old concept is now a commercially viable product [5]. Figure 11.7 shows an induction heater concept powered with a DC coil. An HTS coil is employed as the DC coil in the induction heaters for heating aluminum or copper extrusion billets. The potential for efficiency improvements is substantial because conventional copper coil induction heaters rated for up to around 1 MW operate with an overall efficiency of typically only 50% to 60%. The efficiency of a HTS induction heater expected to be significantly higher than this.

An HTS induction heaters manufactured by Zenergy [5] and its technology partner began operation in 2008 and has revolutionized both energy efficiency and process flexibility in industrial aluminum, brass, bronze, and copper processing. In a precision heating process, HTS induction heaters soften the raw material billets of nonferrous metal in order to improve their ductility. The benefits of the HTS induction heater include the following:

- Energy efficiency over 90% as opposed to 40% to 50% with conventional systems.
- Electromagnetic induction of eddy currents, thus providing homogeneous and precisely controllable temperature levels throughout the material.
- No risk of material damage from local overheating.
- Very short startup time, with improved productivity as a result.
- Highly flexible and cost effective for just-in-time production and manufacturing of high-grade, special-alloy components.
- Economically favorable from day one. Energy cost savings alone amortize the purchase price within five years of operation.

The HTS induction heaters are now a commercial product.

11.3.3 Synchrotron

Synchrotron storage rings use a number of very large copper electromagnets, consuming millions of dollars of electricity per year and requiring substantial amounts of cooling water. HTS magnets can be employed to provide significant energy savings: modeling has indicated a factor of 20 reduction for complete systemwide installations, with equally impressive savings in cooling water demand. The very high current density of HTS wire compared with copper allows more compact coil geometries, leading to greater design flexibility for the magnet and delivering greater optical access to the magnet working area. The HTS magnet advantage is available not just for new facilities: in many cases existing copper coils can be retrofitted with HTS coils without modification to the iron yoke.

The world's first synchrotron magnet (Figure 11.8) fitted with HTS coils was shipped in 2009 from HTS-110, New Zealand, to the Brookhaven National Laboratory in New York. This HTS magnet uses less than half the energy of a copper equivalent, along with substantially less cooling water. Currently the copper coils consume 15 kW of

Figure 11.8 Synchrotron magnet employing an HTS coil (Courtesy of Brookhaven National Lab and HTS-110, NZ)

electricity and significant amounts of cooling water during operation. With each synchrotron operating 50 or more magnets, the energy usage for an entire copper ring is up to 1 MW.

11.4 SUMMARY

The technology for building HTS magnets has been amply tested. These magnets can perform the intended function very efficiently in many industrial and research facilities. However, their initial cost, driven by the high cost of HTS and the cooling system, is inhibiting their wider adaptation. More real-life applications will emerge once these magnets meet the economic goals of their users.

REFERENCES

1. G. Snitchler, S. S. Kalsi, M. Manlief, R. E. Schwall, A. Sidi-Yekhief, S. Ige, R. Medeiros, T. L. Francavilla, and D. U. Gubser, "High-Field Warm-Bore

HTS Conduction Cooled Magnet," *IEEE Trans. Appl. Superconductivity* 9(2, Pt. 1): 553–558, 1999. DOI 10.1109/77.783356.

2. S. S. Kalsi, A. Szczepanowski, G. Snitchler, H. Picard, A. Sidi-Yekhlef, B. Connor, R. E. Schwall, B. MacKinnon, J. L. Tallon, G. Todd, and R. Neale, "Ion Beam Switching Magnet Employing HTS Coils," *IEEE Trans. Appl. Superconductivity* 8(1): 30–33, 1998. DOI 10.1109/77.662692.

3. K. Ohkura, T. Okazaki, and K. Sato, "Large HTS Magnet Made by Improved DI-BSCCO Tapes," *IEEE Trans. Appl. Superconductivity* 18(2): 556–559, 2008. DOI 10.1109/TASC.2008.920807.

4. Qiuliang Wang, Yingming Dai, Xinning Hu, Shouseng Song, Yuanzhong Lei, Chuan He, and Luguang Yan, "Development of GM Cryocooler-Cooled Bi2223 High Temperature Superconducting Magnetic Separator," *IEEE Trans. Appl. Superconductivity* 17(2, Pt. 2): 2185–2188, DOI 10.1109/TASC.2007.899098.

5. L. Masur, C. Buehrer, H. Hagemann, and W. Witte, "Magnetic Billet Heating," *Light Metal Age* (April): 50–55, 2009.

6. Brookhaven National Lab. light source magnet built by HTS-110 of Lower Hutt, New Zealand.

7. R. Gupta and W. Sampson, "Medium and Low Field HTS Magnets for Particle Accelerators and Beam Lines," *IEEE Trans. Appl. Superconductivity* 19(3), 2009, pp 1095–1099.

8. S. S. Kalsi, D. Aized, B. Conner, G. Snitchier, J. Campbell, R. E. Schwall, J. Kellers, T. Stephanblome, A. Tromm, and P. Winn, "HTS SMES Magnet Design and Test Results," *IEEE Trans. Appl. Superconductivity* 7(2, Pt. 1): 971–976, 1992. DOI 10.1109/77.614667.

9. L. Xiao, Z. Wang, S. Dai, J. Zhang, D. Zhang, Z. Gao, N. Song, F. Zhang, X. Xu and L. Lin, "Fabrication and Tests of a 1 MJ HTS Magnet for SMES," *IEEE Trans. Appl. Superconductivity* 18(2): 770–773, 2008. DOI 10.1109/TASC.2008.922234.

10. O. O. Ige, D. Aized, A. Curda, R. Medeiros, C. Prum, P. Hwang, G. Naumovich and E. M. Golda, "Mine Countermeasures HTS Magnet," *IEEE Trans. Appl. Superconductivity* 13(2, Pt. 2): 1628–1631, 2003. DOI 10.1109/TASC.2003.812811.

11. O. O. Ige, D. Aized, A. Curda, D. Johnson and M. Golda, "Test Results of a Demonstration HTS Magnet for Minesweeping," *IEEE Trans. Appl. Superconductivity* 11(1, Pt. 2): 2527–2530, 2001. DOI 10.1109/77.920380.

12. L. Masur, C. Buehrer, H. Hagemann, B. Ostemeyer, and W. Witte, "Magnetic Billet Heating," *Light Metal Age*, April 2009, pp 50–53.

ABOUT THE AUTHOR

Swarn Kalsi received his Ph.D. in electrical engineering from the Imperial College of Science and Technology, London in 1970. He obtained B.Tech. (Hons) and M.Sc. degrees in electrical engineering from the Indian Institute of Technology, Kharagpur in 1962 and Benaras Hindu University, Varanasi, India in 1963, respectively. He started working with the General Electric Global Research, Schenectady, NY in 1971 in the area of applications of superconductors to electric rotating machines and power cables. For almost four decades Dr. Kalsi has remained active in the applications of both low temperature superconductors (LTS) and high temperature superconductors (HTS) to a variety of power equipment, including motors and generators, transformers, fault current limiters, power cables, magnets for fusion and accelerators, and Maglev trains. He has also developed the design features and analysis approaches for such equipment and studied their performance while operating these superconductor devices in an electric grid or on an electric ship.

Dr. Kalsi served as a program manager on numerous programs (both civil and defense) since the early 1970s. He has contributed chapters for many books and has published extensively in IEE, IEEE, Cigre, and other societies. He has served on several IEEE and Cigre panels for work relating to conventional and superconductor technologies. He holds more than 30 US patents.

Currently Dr. Kalsi is retired and works as a consultant for many US and foreign companies.

Applications of High Temperature Superconductors to Electric Power Equipment, by Swarn Singh Kalsi
Copyright © 2011 Institute of Electrical and Electronics Engineers

INDEX

Applications of High Temperature Superconductors to Electric Power Equipment,
by Swarn Singh Kalsi
Copyright © 2011 Institute of Electrical and Electronics Engineers

Printed and bound by CPI Group (UK) Ltd, Croydon, CR0 4YY

16/04/2025

14658604-0002